SEMA SIMAI Springer Series

ICIAM 2019 SEMA SIMAI Springer Series

Volume 3

This sub-series of the SEMA SIMAI Springer Series aims to publish some of the most relevant results presented at the ICIAM 2019 conference held in Valencia in July 2019.

The sub-series is managed by an independent Editorial Board, and will include peer-reviewed content only, including the Invited Speakers volume as well as books resulting from mini-symposia and collateral workshops.

The series is aimed at providing useful reference material to academic and researchers at an international level.

More information about this subseries at http://www.springer.com/series/16499

Ralf Deiterding • Margarete Oliveira Domingues •
Kai Schneider
Editors

Cartesian CFD Methods for Complex Applications

Springer

Editors
Ralf Deiterding
Department of Aeronautics & Astronautics
University of Southampton
Southampton, UK

Margarete Oliveira Domingues
Associate Laboratory of Computing and Applied Mathematics
National Institute for Space Research (INPE)
São José dos Campos
Brazil

Kai Schneider
Institut de Mathématiques de Marseille (I2M)
Aix-Marseille Université
Marseille, France

ISSN 2199-3041 ISSN 2199-305X (electronic)
SEMA SIMAI Springer Series
ISSN 2662-7183 ISSN 2662-7191 (electronic)
ICIAM 2019 SEMA SIMAI Springer Series
ISBN 978-3-030-61760-8 ISBN 978-3-030-61761-5 (eBook)
https://doi.org/10.1007/978-3-030-61761-5

This Springer imprint is published by the registered company Springer Nature Switzerland AG.
The registered company address is: Gewerbestrasse 11, 6330 Cham, Switzerland

Preface

Cartesian discretization approaches are ubiquitous in computational fluid dynamics. When applied to problems in geometrically complex domains or fluid–structure coupling problems, Cartesian schemes allow for automatic and scalable meshing; however, order-consistent immersed boundary conditions and efficient dynamic mesh adaptation take forefront roles. This volume contains selected contributions from the four-session thematic mini-symposium on "Cartesian CFD Methods for Complex Applications" at ICIAM 2019 held in Valencia in July. The papers highlight cutting-edge applications of Cartesian CFD methods and describe the employed algorithms and numerical schemes. An emphasis is laid on complex multi-physics applications such as magnetohydrodynamics or aerodynamics with fluid–structure interaction, solved with various discretizations, e.g. finite difference, finite volume, multi-resolution or lattice Boltzmann CFD schemes. Software design and parallelization challenges are also addressed briefly.

The volume is organized into two parts of three contributions each. Part one is focused on incompressible flows and has the following contributions: Bergmann et al. propose an adaptive finite-volume method with quad-tree discretization of the incompressible Navier–Stokes equations. Moving immersed bodies are modelled with volume penalization, and their interface is tracked using level sets. Test cases with flows around cylinders show the validity and precision of the approach. Fluid–structure interaction for flexible insect wings is studied in the paper by Truong et al. A mass spring model is used for the wing structure. The fluid solver is based on a Fourier pseudospectral discretization with volume penalization to take into account the complex and time-varying geometry. Applications consider flapping bumblebee flight in laminar and turbulent flow. The paper by Kadri and Perrier presents a numerical scheme for incompressible Navier–Stokes equations in three dimensions using divergence-free wavelets. Constructions for these basis functions are given for no-slip and free-slip boundary conditions and divergence-free wavelets in dimension higher than three are given. Numerical examples illustrate the scheme for lid-driven cavity problems.

The second part deals with compressible and weakly compressible flows and has likewise three contributions. Perron et al. propose an immersed boundary method for compressible flows using structured Cartesian grids. A direct forcing approach based on the use of ghost cells is chosen. Two flow configurations are considered, a supersonic flow around a blunt body to demonstrate the capability of mesh adaptation to increase the accuracy and a large eddy simulation of the flow around a three-dimensional high-lift airfoil. Comparisons with experimental data and a reference body-fitted computation are as well presented. Moreira Lopes et al. discuss the performance and detail verification and validation of a wavelet-adaptive magnetohydrodynamic solver, realized within the MPI-parallel AMROC (Adaptive Mesh Refinement in Object-oriented C++) framework. A prototype simulation fuses this solver with actual satellite date for space weather forecasting. Finally, Gkoudesnes and Deiterding report on the incorporation of the lattice Boltzmann method into the AMROC environment. The algorithmic details and verification of large eddy simulation with the wall-adapting local eddy-viscosity model for dynamically adapting meshes and with ghost cell-based embedded boundary conditions are presented.

We thank all the speakers of the four sessions for making this mini-symposium a successful event, and we are grateful to the authors for their contributions. We are indebted to the numerous referees for their constructive and detailed reports. For all papers, we had three to four reviews, improving thus further the quality of this edited volume.

Southampton, UK
São José dos Campos, Brazil
Marseille, France
May 2020

Ralf Deiterding
Margarete Oliveira Domingues
Kai Schneider

Contents

About the Editors

Ralf Deiterding is currently Associate Professor in Fluid Dynamics at the University of Southampton. He graduated with a Master's in Technomathematics from the Technical University of Clausthal and has obtained a PhD in Applied Mathematics and CFD from the Technical University Cottbus. His research focuses on the development and application of innovative high-resolution and multi-scale simulation methods for CFD. He is the main author of the simulation frameworks AMROC and Virtual Test Facility.

Margarete Oliveira Domingues is senior researcher at the National Institute for Space Research (INPE), Brazil. She obtained her Master's in Meteorology at INPE, and PhD degree in Applied Mathematics from the University of Campinas (UNICAMP), Brazil. Her research activities are focused on nonlinear analysis, multi-scale techniques and wavelets for scientific computing and their application to space flows and data, including space magnetohydrodynamic developments and multi-dimensional signal processing.

Kai Schneider is Professor of Mechanics and Applied Mathematics at the Aix-Marseille University, Marseille, France. He obtained his Master's and PhD degree from the University of Kaiserslautern, Germany, and his habilitation from the University Strasbourg, France. His research activities are focused on multi-scale techniques and wavelets for scientific computing and their application to turbulent flows, including fluid–structure interaction, e.g. for insect flight, and magnetohydrodynamic turbulence.

AMR Enabled Quadtree Discretization of Incompressible Navier–Stokes Equations with Moving Boundaries

Michel Bergmann, Antoine Fondanèche, and Angelo Iollo

Abstract We present a versatile finite-volume method for the simulation of incompressible flows past moving bodies. The Navier–Stokes equations are discretized on AMR enabled quadtree grids, where the dynamic in time refinement is adapted to the evolution of the fluid–solid system. The immersed bodies are modeled through a second-order volume penalization method, and the interface is tracked using a level-set description. We highlight on two dimensional test cases that the uniform grids accuracy can be recovered using quadtree grids with less degrees of freedom.

1 Introduction

Efficient numerical tools to simulate fluid–solid interactions are in interest in a wide range of application fields, from engineering to medical applications. For instance, the simulation of a flow around a wind turbine blade [1] or in cardiac support devices [2] is an essential support to optimize the design of these new technologies.

To face this challenge, a large number of studies have been carried out to precisely describe these interactions, especially when dealing with complex geometries. These studies are based on two numerical approaches. The first approach is based on the Arbitrary Lagrangian–Eulerian methodology for which flows are calculated on a moving mesh in a time-varying area (see [3] for details). These methods are generally very accurate, based on sophisticated numerical schemes, but are difficult to implement, especially for the consideration of structures with large deformations. The generation of a body-fitted mesh is expensive, and the use of a dynamic mesh partitioner for parallel calculations is moreover necessary. The second approach is based on fictitious domain methods, such as immersed boundary

M. Bergmann · A. Fondanèche (✉) · A. Iollo
Equipe-Projet Memphis, INRIA Bordeaux-Sud Ouest, Talence, France

Université de Bordeaux, IMB, UMR 5251, Talence, France
e-mail: michel.bergmann@inria.fr; antoine.fondaneche@inria.fr; angelo.iollo@inria.fr

R. Deiterding et al. (eds.), *Cartesian CFD Methods for Complex Applications*, SEMA SIMAI Springer Series 3, https://doi.org/10.1007/978-3-030-61761-5_1

methods [4] or augmented Lagrangian approaches [5], which represent a balance between precision and practicability of the simulation.

In this work, we use the Brinkman penalization method [6] that is an embedded interface method such as the immersed boundary method (IBM, introduced by Peskin [7, 8]) with discrete forcing. In the context of interface-capturing methods for simulating multiphase flows, such as volume-of-fluid [9], phase field [10], or level-set [11] descriptions, the whole system is strongly coupled as soon as both materials are subject to the same constitutive equation. Here we consider a level-set formulation with the sign distance function, where the fluid–solid interface is defined by the zero isoline. Cartesian methods for incompressible flows [12–14] need a very refined mesh to get accurate results because they need a good representation of the body geometry. With respect to these methods, we propose a quadtree-based method that provides an equivalent accuracy with a smaller number of grid points. By refining the mesh in regions of interest, such as in the vicinity of the interface or where the solution varies significantly and by coarsening where the solution varies slightly, the computational time is significantly decreased with a limited loss of accuracy.

2 The Penalized Navier–Stokes Model

The aim of this work is to study the interaction between an incompressible Newtonian fluid and some rigid moving bodies. A square domain $\Omega \subset \mathbb{R}^2$ is decomposed into two parts, namely $\Omega = \Omega_f \cup \Omega_s$, where Ω_f and Ω_s denote the fluid and solid domains, respectively. The fluid–solid interface is $\Gamma(t) = \partial\Omega_s(t)$. The sketch of the flow configuration is presented in Fig. 1. The governing equations are

$$\frac{\partial \mathbf{u}}{\partial t} + (\mathbf{u} \cdot \nabla)\mathbf{u} = -\frac{1}{\rho}\nabla p + \nu\Delta\mathbf{u} \qquad \text{in } \Omega_f, \tag{1a}$$

$$\nabla \cdot \mathbf{u} = 0 \qquad \text{in } \Omega_f, \tag{1b}$$

$$\mathbf{u} = \mathbf{u}_s \qquad \text{on } \Gamma(t), \tag{1c}$$

$$\mathbf{u}(t = 0, \cdot) = \mathbf{u}_0 \qquad \text{in } \Omega_f, \tag{1d}$$

where $\mathbf{u} = (u, v)^T$ is the velocity field, p is the pressure, ρ is the density, and ν is the kinematic viscosity of the fluid. Finally, $\mathbf{u}_0$ is the initial condition and $\mathbf{u}_s$ is the velocity of the body interface.

The volume penalization approach introduced in [6] is chosen. The main idea of this method is to consider the whole system as porous media, with a variable permeability ε. The solid structure is considered to have a very low permeability

Fig. 1 Sketch of the flow setup

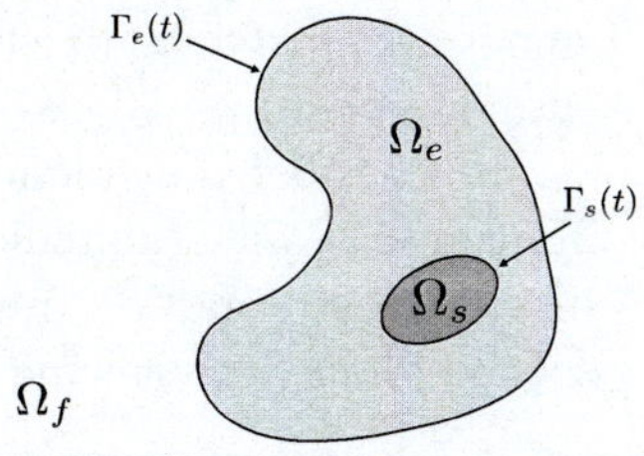

$\varepsilon \ll 1$. The Navier–Stokes equations (1a), (1b), and (1c) can thus be solved in a coupled way in Ω as

$$\frac{\partial \mathbf{u}}{\partial t} + (\mathbf{u} \cdot \nabla)\mathbf{u} = -\frac{1}{\rho}\nabla p + \nu \Delta \mathbf{u} + \frac{\chi_s}{\varepsilon}(\mathbf{u}_s - \mathbf{u}) \tag{2a}$$

$$\nabla \cdot \mathbf{u} = 0, \tag{2b}$$

where χ_s is the characteristic function of the solid, defined as

$$\chi_s(\mathbf{x}) = \begin{cases} 1 & \text{if } \mathbf{x} \in \overline{\Omega}_s \\ 0 & \text{if } \mathbf{x} \in \Omega_f. \end{cases}$$

The position of the solid Ω_s is tracked with a level-set function ϕ, chosen as the signed distance to the interface $\Gamma(t)$ with a negative sign inside the solid and a positive sign inside the fluid. The interface is then defined as the zero isoline of ϕ, namely, $\Gamma(t) = \{\mathbf{x} \in \mathbb{R}^2 : \phi(x) = 0\}$. As a consequence, the characteristic function χ_s may be defined using the level-set function as $\chi_s(\mathbf{x}) = 1 - H(\phi(\mathbf{x}))$, where H is the Heaviside function. Using this method, both solid and fluid equations are solved, with no distinction. The solution of the system (2) converges towards the solution of the decoupled system (1) (see [6]) as $\sqrt{\varepsilon}$ tends to zero [15]. In practice, the choice $\varepsilon = 10^{-10}$ is suitable for all our simulations.

3 Discretization of the Governing Equations

3.1 *Time Integration*

We denote by Δt the time step such that $t^{n+1} = t^n + \Delta t$ and $\varphi^n := \varphi(t^n)$ the discrete value of a function φ at the time t^n. For the sake of simplicity, the time step is assumed to be fixed here, but the generalization to an adaptative time step is straightforward.

We consider the fractional time step method introduced by Chorin [16] and Temam [17]. First, a prediction step is performed to get a preliminary estimate $\mathbf{u}^*$ of the velocity starting from a guess for the pressure field q. This is done using a second-order Gear scheme as

$$\frac{3\mathbf{u}^* - 4\mathbf{u}^n + \mathbf{u}^{n-1}}{2\Delta t} + \left(2\nabla\cdot(\mathbf{u}\otimes\mathbf{u})^n - \nabla\cdot(\mathbf{u}\otimes\mathbf{u})^{n-1}\right) = \frac{1}{\rho}\Big(- \nabla q + \mu\Delta\mathbf{u}^* + \frac{\chi_s^{n+1}}{\epsilon}(\mathbf{u}_s^{n+1} - \mathbf{u}^*)\Big). \tag{3}$$

We use here the incremental version of the projection method proposed by Goda [18] for which $q := p^n$. A second-order accurate volume penalization method is employed as in [12], for which the velocity inside the body $\mathbf{u}_s$ is artificially imposed by image point correction (IPC). As long as the interface does not fit the grid points, this technique ensures that the velocity of the interface $\mathbf{u}|_{\partial\Omega_s}$ is enforced, and therefore, the velocity gradient in the first layer of fluid is consistent.

Since the predicted velocity field is not divergence free, a projection step in a solenoidal subspace is performed

$$\frac{\mathbf{u}^{n+1} - \mathbf{u}^*}{\Delta t} = -\frac{1}{\rho}\left(\nabla p^{n+1} - \nabla q\right) \quad \text{in } \Omega. \tag{4}$$

Since we want to recover a divergence-free velocity, i.e., $\nabla\cdot\mathbf{u}^{n+1} = 0$, by applying the divergence operator to Eq. (4), we get

$$\Delta p' = \nabla\cdot\mathbf{u}^* \quad \text{in } \Omega, \tag{5}$$

where we denote by $p' := \frac{\Delta t}{\rho}(p^{n+1} - q)$ the pressure increment. Homogenous Neumann boundary conditions are imposed in order to ensure that there is no perturbation at the boundaries for the normal velocity.

As soon as the increment of pressure p' is determined by solving the Poisson equation (5), the pressure p^{n+1} can be updated and the velocity field is corrected as follows:

$$p^{n+1} = q + \frac{\rho}{\Delta t}\, p', \tag{6a}$$

$$\mathbf{u}^{n+1} = \mathbf{u}^* - \nabla p'. \tag{6b}$$

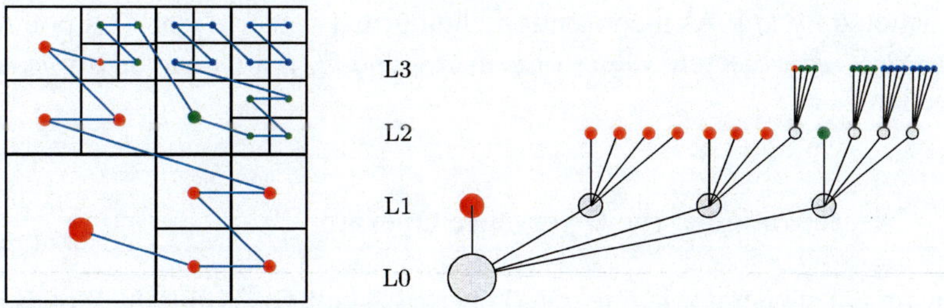

Fig. 2 Graded quadtree grid with the global Z-ordering. The different colors depicts the balancing between processors

3.2 Spatial Discretizations

The computational domain Ω is discretized with a quadtree grid. As depicted in Fig. 2, a quadtree grid is composed of square cells with different levels of refinement. Here, the hierarchical grid is graded, which means that the difference of level between a cell and all its adjacent cells (called *neighbors*) is at most one. Thanks to the library PABLO, as a part of Bitpit library,[1] we get use of an efficient tool for storing the data structure. Following the linear Z-ordering proposed by Morton in 1966 [19], we can get access to data coming from neighboring cells in an optimized way from computational cost and memory aspects. Moreover, Adaptative Mesh Refinement (AMR) is used to adapt the mesh dynamically to the flow configuration by refining in the areas of interest, such as wakes of bodies, vortices, or around the structures, which is even more interesting when the structures can move or be deformed. For the domain decomposition, the number of communications between processors is limited to only one layer of ghost cells. While this constraint guarantees a very high scalability of the parallelism, the discretizations of the operators are built with compact stencils limiting the order of numerical scheme.

In this section, we detail the finite-volume discretizations of the operators involved. To describe these discretizations, let φ be a scalar function and $\mathbf{v}$ a vector field. The square domain Ω is decomposed into a quadtree partition of N_{cells} square cells Ω_i of level L_i (being the *leaves* of the tree) such that $\Omega = \dot{\bigcup}_i \Omega_i$. By convention, the grid configuration is identified by its minimum and maximum levels of refinement L_{min} and L_{max}. In other words, for a $L_{min} - L_{max}$ grid, the characteristic length h_i of Ω_i is between $h_{min} = \min_k h_k$ and $h_{max} = \max_k h_k$. A two dimensional uniform L^{ℓ} grid is hence composed of $2^{2\ell}$ cells. We denote by $\mathbf{x}_i$ the center of the cell Ω_i, $|\Omega_i|$ its area, and $\varphi_i := \varphi(\mathbf{x}_i)$ the discrete value of a quantity φ evaluated at the cell center $\mathbf{x}_i$.

For a finite-volume method, the discrete operators are computed as face contributions called fluxes. Let f be the intersecting face of Ω_{out} and Ω_{in}. The length of

[1] https://optimad.github.io/bitpit.

f is denoted by $|f|$. As a convention, the normal vector $\mathbf{n}_f$ of f is pointing from Ω_{in} to Ω_{out}. The discrete values of φ in Ω_{in} and Ω_{out} are denoted by φ_{in} and φ_{out}, respectively.

3.2.1 Discretization of the Divergence Operator

The divergence operator is integrated over each cell Ω_i. Using the Stokes theorem, the volume integral is transformed into a surface integral as

$$\nabla \cdot \mathbf{v}\big|_{\Omega_i} = \frac{1}{|\Omega_i|} \int_{\Omega_i} \nabla \cdot \mathbf{v}\, d\mathbf{x} = \frac{1}{|\Omega_i|} \oint_{\partial\Omega_i} \mathbf{v} \cdot \mathbf{n}\, ds, \tag{7}$$

where $\mathbf{n}$ is the outward normal vector of the boundary $\partial\Omega_i$. By decomposing the whole boundary into separate faces, the discrete value of $\nabla \cdot \mathbf{v}$ on Ω_i can be computed as

$$(\nabla \cdot \mathbf{v})_i = \frac{1}{|\Omega_i|} \sum_{f \subset \partial\Omega_i} \mathbf{v}_{fc} \cdot \mathbf{n}_f \, |f|, \tag{8}$$

where subscript fc refers to the center position of the face f. Using the relation $\nabla\varphi = \nabla \cdot (\varphi I)$, the discrete cell-center gradient $(\nabla\varphi)_i$ is estimated similarly.

If the collocated cell-center velocity $\mathbf{u}^*$ is used to compute $\nabla \cdot \mathbf{u}^*$ in Eq. (5), some spurious grid-to-grid oscillations may appear due to odd–even decoupling between velocity and pressure. This is one of the main drawbacks for non-staggered grids. As a consequence, this decoupling causes large variations of pressure that are even more critical for quadtree grids, and this problem can lead to numerical instabilities at the level jumps. As shown by Ferziger and Peric [20], traditional collocated methods cannot guarantee the pressure smoothness and the mass conservation simultaneously. One way to overcome this problem has been proposed by Patankar [21] and consists in a fully staggered arrangement of the variables $(\mathbf{u}, p)$. For this kind of methods, the prediction step (3) and the Poisson equation (5) are solved at different locations, which leads to different spatial discretizations. In this sense, staggered arrangements become more challenging for Cartesian methods.

In order to stabilize the method, the collocated approach introduced by Rhie and Chow [22] for steady flows, and Zang et al. [23] for unsteady flows, is considered. A face-center velocity called U^*_{fc} is defined in Ω (see [13]) as

$$\tilde{\mathbf{u}} = \mathbf{u}^* + \frac{\Delta t}{\rho} (\nabla p^n)_{cc} \tag{9a}$$

$$\widetilde{U_{fc}} = \mathcal{F}(\tilde{\mathbf{u}}) \tag{9b}$$

$$U^*_{fc} = \widetilde{U_{fc}} - \frac{\Delta t}{\rho} (\nabla p^n)_{fc}, \tag{9c}$$

Fig. 3 Pattern of the arrangement of the cell-center (red squares) and face-center (blue dots) velocities on quadtree grids

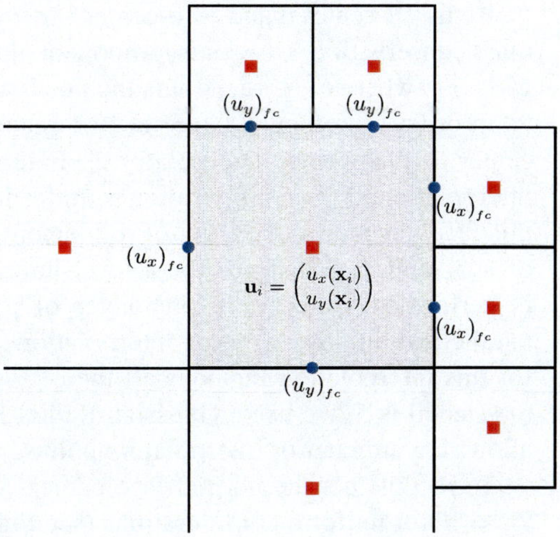

where subscripts cc and fc refer to cell-center and face-center locations, respectively, and $\mathcal{F}$ is the operator used for the interpolation of the normal face-center velocity, using the cell-center velocity. The operator $\mathcal{F}$ is detailed at the end of the section. This velocity arrangement is illustrated in Fig. 3.
The face-center velocity U^*_{fc} is now used in the Poisson equation (5) that becomes

$$\Delta p' = \nabla \cdot U^*_{fc}. \tag{10}$$

Once the pressure increment p' is obtained, both cell-center and face-center velocities are finally corrected as follows:

$$\mathbf{u}^{n+1} = \mathbf{u}^* - (\nabla p')_{cc} \tag{11a}$$

$$U^{n+1}_{fc} = U^*_{fc} - (\nabla p')_{fc}. \tag{11b}$$

If the face-center velocity U^{n+1}_{fc} satisfies the discrete mass conservation within the limit of the Poisson solver tolerance ϵ, i.e.,

$$\sum_{f \subset \partial\Omega_i} U^{n+1}_{fc} |f| = \mathcal{O}(\epsilon) \ll 1, \quad \forall \Omega_i \subset \Omega,$$

$\mathcal{F}(\mathbf{u}^{n+1})$ does not. As a consequence, the use of U^{n+1}_{fc} is hence necessary for the computation of the convective term, see Sect. 3.2.3.

Both cell-center $\mathbf{u}$ and face-center U_{fc} velocities need thus to be stored to ensure mass conservation. Only one component of the face-center velocity is needed, being $U_{fc} \cdot \mathbf{e}_f$, where $\mathbf{e}_f \in \{\mathbf{e}_x, \mathbf{e}_y\}$ is the positive unitary normal vector of the face f.

In order to compute a face-center quantity φ_{fc}, we adopt a second-order cell-center to face-center interpolator $\mathcal{I}$. Figure 4 illustrates the stencil used for this interpolation. If the configuration is uniform, which means that the two cells sharing this face have the same level of refinement, φ_{fc} is basically chosen as the average of φ_{out} and φ_{in}. Instead, for a level-jump configuration, a bilinear interpolation is performed. Since only one layer of ghost cells is available because of the parallelization, higher order interpolations are unsuitable. The stencil considered for this interpolation includes all the cells neighboring the face. Since the size of the stencil is 5, we use a Gaussian Radial Basis Function (RBF) interpolation that allows any number of interpolating points.

To reconstruct the normal face-center velocity $\widetilde{U}_{fc}$, it is appropriate to choose $\mathcal{F} := \mathcal{I}$ for uniform configurations (see Fig. 4, left). However, close to level jumps (see Fig. 4, right), this direct interpolation introduces spurious oscillations due to an inaccurate computation of $\nabla \cdot U^*_{fc}$. To limit this loss of accuracy, the face-center velocity is calculated using U^n_{fc} as

$$\widetilde{U}_{fc} = U^n_{fc} + \mathcal{I}(\tilde{\mathbf{u}} - \mathbf{u}^n). \tag{12}$$

This stabilization ensures that the solution remains smooth near level jumps.

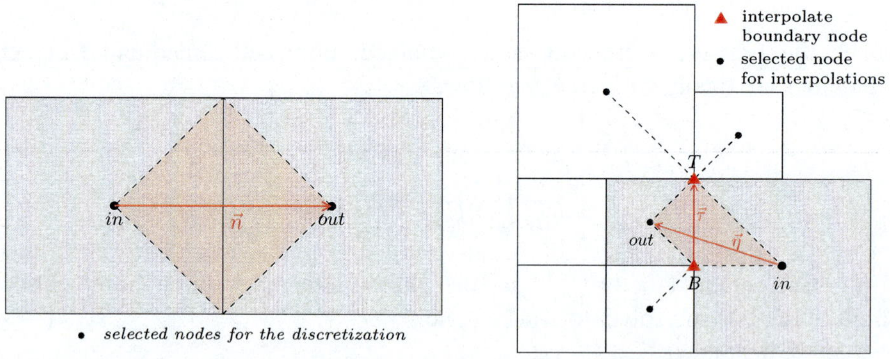

Fig. 4 Pattern of the diamond FV method. The red quadrilateral represents the dual cell used to interpolate $(\nabla\varphi)_{fc}$, and the black dots refer to the stencil used for the interpolation of φ_{fc}. Left: uniform configuration, right: level-jump configuration

3.2.2 Discretization of the Laplacian Operator

The Laplacian operator is discretized using a diamond finite-volume method [24, 25]. Integrating $\Delta\varphi$ on a cell Ω_i and applying Stokes theorem yield

$$(\Delta\varphi)_i = \frac{1}{|\Omega_i|}\int_{\Omega_i} \Delta\varphi \, d\mathbf{x} = \frac{1}{|\Omega_i|}\int_{\partial\Omega_i} \nabla\varphi \cdot \mathbf{n} \, ds \tag{13}$$

where $\mathbf{n}$ is the normal vector of $\partial\Omega_i$. The diffusion term is then approximated as

$$(\Delta\varphi)_i = \frac{1}{|\Omega_i|}\sum_{f\subset\partial\Omega_i} (\nabla\varphi)_{fc} \cdot \mathbf{n}_f |f|. \tag{14}$$

We denote the endpoints of each face f by T for top and B for bottom, as depicted in Fig. 4. The normal velocity gradient on this face is computed as the solution of the following system:

$$\nabla\varphi \cdot \boldsymbol{\eta} = \varphi_{out} - \varphi_{in} \tag{15a}$$

$$\nabla\varphi \cdot \boldsymbol{\tau} = \varphi_T - \varphi_B \tag{15b}$$

with

$$\boldsymbol{\eta} = \mathbf{x}_{out} - \mathbf{x}_{in} \quad \text{and } \boldsymbol{\tau} = \mathbf{x}_T - \mathbf{x}_B.$$

The simplest case occurs when Ω_{out} and Ω_{in} have the same level of refinement, namely, $|\Omega_{out}| = |\Omega_{in}|$. Here, $\boldsymbol{\eta} = \mathbf{n}$; thus, Eq. (15a) is a classical second-order centered difference approximation of the normal gradient. If Ω_{in} and Ω_{out} have different levels of refinement, φ_T and φ_B have to be interpolated. According to the configuration, a $\mathbf{P}_1$ (3-point linear) or $\mathbf{Q}_1$ (4-point bilinear) interpolation is considered, depending on the number of grid cells that the node belongs to.
Let us describe the $\mathbf{Q}_1$ interpolation. The value of φ at the point $\mathbf{x}$ can be described through a bilinear basis as

$$\varphi(\mathbf{x}) = c_1 + c_2 x + c_3 y + c_4 xy, \tag{16}$$

where the coefficients c_i have to be determined. They are calculated by solving the linear system

$$\begin{pmatrix} 1 & x_1 & y_1 & x_1y_1 \\ 1 & x_2 & y_2 & x_2y_2 \\ 1 & x_3 & y_3 & x_3y_3 \\ 1 & x_4 & y_4 & x_4y_4 \end{pmatrix} \begin{pmatrix} c_1 \\ c_2 \\ c_3 \\ c_4 \end{pmatrix} = \begin{pmatrix} \varphi(\mathbf{x}_1) \\ \varphi(\mathbf{x}_2) \\ \varphi(\mathbf{x}_3) \\ \varphi(\mathbf{x}_4), \end{pmatrix} \tag{17}$$

which comes from the evaluation of Eq. (16), at the 4 points previously defined. Finally, using the coefficients $c_i = c_i\big(\varphi(\mathbf{x}_1), \varphi(\mathbf{x}_2), \varphi(\mathbf{x}_3), \varphi(\mathbf{x}_4)\big)$, the values $\varphi(\mathbf{x}_T)$ and $\varphi(\mathbf{x}_B)$ are then recovered by assessment of Eq. (16) at the points $\mathbf{x}_T$ and $\mathbf{x}_B$. Since the normal vector $\mathbf{n}_f$ can be expressed as a linear combination of the two previously defined vectors $\boldsymbol{\eta}$ and $\boldsymbol{\tau}$, i.e.,

$$\exists c_\tau, c_\eta \in \mathbb{R} : \ \mathbf{n}_f = c_\tau \boldsymbol{\tau} + c_\eta \boldsymbol{\eta},$$

the normal gradient at the face f is approximated as a linear combination with the whole compact stencil of the face named $\mathcal{N}_f$ as

$$\nabla\varphi \cdot \mathbf{n}_f = c_\tau(\varphi_T - \varphi_B) + c_\eta(\varphi_{out} - \varphi_{in}) \equiv \sum_{i \in \mathcal{N}_f} \omega_i \varphi_i. \tag{18}$$

3.2.3 Discretization of the Convective Term

The numerical computation of the convective term on each cell Ω_i

$$\mathbf{C}_i = \frac{1}{|\Omega_i|}\int_{\Omega_i} \nabla\cdot(\mathbf{u}\otimes\mathbf{u})\, d\mathbf{x} = \frac{1}{|\Omega_i|}\int_{\Omega_i} \mathbf{u}\cdot\nabla\mathbf{u}\, d\mathbf{x} = \frac{1}{|\Omega_i|}\int_{\partial\Omega_i} \mathbf{u}(\mathbf{u}\cdot\mathbf{n})\, ds \tag{19}$$

is performed using an upwind scheme. Since we use the fact that the velocity is divergence-free in (19), the normal face-center velocity U_{fc} introduced previously in Sect. 3.2.1 is used. The convective term is approximated as

$$\mathbf{C}_i = \frac{1}{|\Omega_i|} \sum_{f \subset \partial\Omega_i} \mathbf{u}_{fc}(U_{fc} \cdot \mathbf{n}_f)|f|. \tag{20}$$

More generally, we describe hereafter the computation of the numerical flux $F(\varphi) = U_{fc}\varphi$ for any scalar function φ.

A linear piecewise polynomial reconstruction is performed as an interpolation $\widetilde{\varphi}$ of φ. In cell Ω_i, the polynomial is defined as

$$\widetilde{\varphi}_i(\mathbf{x}) := (\widetilde{\varphi})_{\Omega_i}(\mathbf{x}) = \varphi_i + (\mathbf{x} - \mathbf{x}_i)^T(\nabla\varphi)_i, \tag{21}$$

where $\varphi_i = \varphi(\mathbf{x}_i)$ and $(\nabla\varphi)_i$ is the discrete value of its gradient, computed as the finite-volume approximation introduced in Sect. 3.2.1. For any point $\mathbf{x} \in f$, in particular for its center $\mathbf{x}_{fc}$, the quantity φ can be interpolated from both sides of the face, namely, $\varphi^- = \widetilde{\varphi}_{in}(\mathbf{x}_{fc})$ and $\varphi^+ = \widetilde{\varphi}_{out}(\mathbf{x}_{fc})$. To guarantee that the discretization is upwind, the Rusanov numerical flux [26] is used:

$$\widetilde{\mathbf{F}}(\varphi_{fc}) := \mathbf{F}_{LLF}(\varphi^-, \varphi^+) = \frac{1}{2}\Big(F(\varphi^+) + F(\varphi^-)\Big) - \frac{\alpha}{2}|U_{fc}|(\varphi^+ - \varphi^-), \tag{22}$$

where the parameter of stabilization α is set to 1.
Special treatments are used on the domain boundaries where ghost cells are created.

4 Numerical Validations

Both static and dynamic quadtree grids are used.The dynamic grids give the ability to ensure a better precision by refining in the areas of interest following a well-defined AMR criterion. To this end, it is possible to make use of different kinds of refinement criteria based for instance on the vorticity, the Q-criterion, or the velocity gradient. In this work, a criterion C based on the Hessian $\mathbf{H}$ of the velocity is used:

$$C = \frac{||\mathbf{H}(\mathbf{u})||_F}{\max_k ||\mathbf{u}_k||_2}, \tag{23}$$

where $|| \cdot ||_F$ and $|| \cdot ||_2$ refer to the Frobenius and Euclidian norms, respectively.

For a $L_{min} - L_{max}$ grid, an appropriate level of refinement is thus determined between L_{min} and L_{max} using some thresholds denoted by C_{min} and C_{max}. According to the computed value of C_i, the optimal level of a cell Ω_i is maximum (resp., minimum) if $C_i > C_{max}$ (resp., $C_i < C_{min}$). In practice, these thresholds C_{min} and C_{max} are fixed to 0.5 and 5, respectively. The mesh adaptation is only performed every n time step, where n is user-defined and depends on the test case considered, or if the interface gets too close to a coarse cell.

In the context of level-set methods, a Lagrangian computation of forces becomes unachievable since the position of the interface is not precisely tracked in time. To deal with this problem, we use the Eulerian approach proposed by Noca [27], known as the control volume method. We introduce a fictive domain $\widetilde{\Omega_s}$ surrounding the solid domain Ω_s, being a rectangular domain that fits exactly the grid cells (the whole volume is hence exactly known). The force can be written as

$$\mathbf{F} = -\frac{d}{dt} \int_{\widetilde{\Omega_s}} \rho \mathbf{u} \, d\mathbf{x} + \int_{\partial \widetilde{\Omega_s}} \big(\mathbb{T}(\mathbf{u}, p) - \rho \mathbf{u}(\mathbf{u} - \mathbf{u}_s) \big).\mathbf{n} \, ds, \tag{24}$$

where $\mathbb{T}(\mathbf{u}, p) = -p\mathbf{I} + \mu(\nabla \mathbf{u} + \nabla \mathbf{u}^T)$ is the stress tensor. The forces are hence approximated using a second-order least-square interpolation.

4.1 Flow Past a Cylinder

In a two dimensional context, a steady cylinder with a diameter $D = 1$, centered at $(0, 0)$, is immersed in a fluid at constant velocity $U_\infty = 1$. The inflow is imposed with Dirichlet boundary conditions $\mathbf{u} = U_\infty \mathbf{e}_x$. At the outflow, a homogeneous Neumann condition is applied. Finally, streamline boundary conditions, $v = 0$ and $\frac{\partial u}{\partial n} = 0$, are chosen on the other boundaries. The computational domain $\Omega = [-8D, 16D] \times [-12D, 12D]$ is quite large to reduce the effect of the boundary conditions. We use here the dimensionless version of the Navier–Stokes equations.

4.1.1 Re = 200

We focus on a long time integration at $Re = 200$. The simulations are run over static quadtree grids, based on a refinement by boxes, as depicted in Fig. 5 (left). The most refined box is chosen in order to obtain a high mesh resolution around the cylinder as well as in the wake direction. After reaching the asymptotic regime, the vortices are well preserved, even in the region where the grid is coarsened, see Fig. 5 (right). The temporal evolution of the drag C_D and lift C_L coefficients is plotted in Fig. 6. To compare with the literature results, we compute the mean drag coefficient and Strouhal number $S_t = fD/U_\infty$, where f is the frequency of the vortex shedding. On uniform grids, these quantities, which are reported in Table 1, are in good agreement with the literature.

We now pay attention on the effect of coarsening on the overall accuracy of the method. Starting from a uniform grid, the computation is also performed on three intermediate quadtree grids with one, two, and three levels of coarsening, respectively. For the four computations, the meshes are locally identical close to the cylinder. As a comparison, the aerodynamic coefficients and the Strouhal number

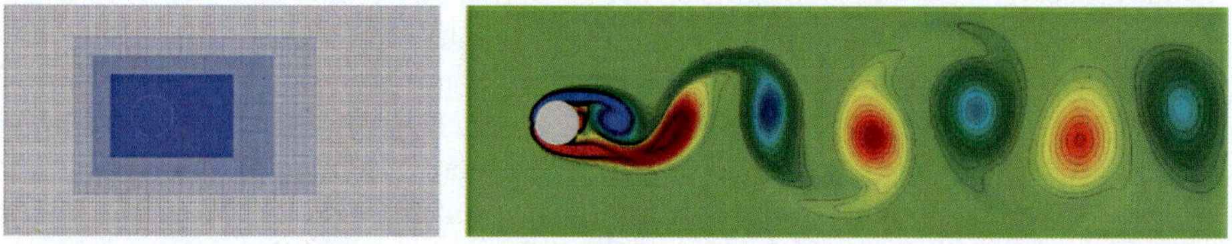

Fig. 5 Simulation on the $L_8 - L_{11}$ static quadtree grids for the flow past a cylinder test at $Re = 200$. Left: quadtree grid employed, right: z-component of the vorticity at $t = 150$ s

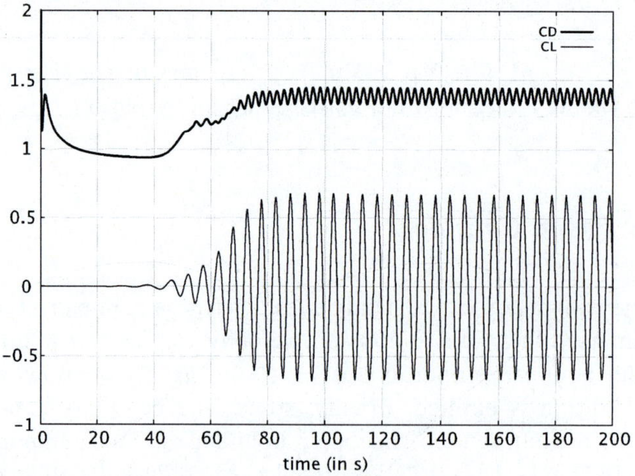

Fig. 6 Aerodynamic coefficient for the flow past a cylinder at $Re = 200$. Results obtained with the L8–L11 quadtree grid

Table 1 Comparison of the mean drag coefficient and the Strouhal number with the previous studies for the flow past a cylinder at $Re = 200$ on uniform grids

Authors	C_D	S_t
Braza et al. [28]	1.4000	0.2000
Henderson [29]	1.3412	0.1971
He et al. [30]	1.3560	0.1978
Bergmann [31]	1.3900	0.1999
Taymans [32]	1.3951	0.2039
Present work	1.3920	0.2013

Table 2 Influence of grid coarsening on the aerodynamic coefficients and Strouhal number for the flow past a cylinder at $Re = 200$. The maximum level refinement is $L_{max} = 11$

Minimum grid level	Number of grid points	C_D	C_L	S_t
11	4,194,304	1.392 ± 0.061	0 ± 0.704	0.2013
10	1,081,600	1.392 ± 0.062	0 ± 0.702	0.2012
9	312,832	1.391 ± 0.061	0 ± 0.694	0.2006
8	125,824	1.369 ± 0.054	0 ± 0.626	0.1956

are given in Table 2. For one or two levels of coarsening, the overall results are really close to those obtained with the uniform one. More precisely, the relative differences on S_t and the amplitude of oscillations are less than 0.4 and 1.4%, respectively, if compared to the uniform grid configuration. With three levels of coarsening, the difference is amplified but is still acceptable. However, the number of grid cells N_{cells} is divided by 13 providing to a good balance between accuracy and computational costs.

4.1.2 Re = 550

We then examine a short time integration $t \in [0, 5]$ for $Re = 550$. The drag coefficient is computed over time and compared to the results provided by Ploumhans [33]. This test case is run over two different flow simulations, namely:

- a steady cylinder is immersed in an infinite flow ($\mathbf{u}_s = \mathbf{0}$, $U_\infty = 1$),
- an impulsively started cylinder moving with a constant velocity 1 is immersed in a fluid at rest ($\mathbf{u}_s = -\mathbf{e}_x$, $U_\infty = 0$).

In these two flow configurations, the forces applied on the body are identical, which leads theoretically to the same drag coefficients. Figure 7 depicts the time evolution of the drag coefficient for both flow configurations.

For the steady cylinder test, the results obtained on static quadtree grids are in good agreement with the literature. From the $L_8 - L_{11}$ configuration, the solution has already converged and is as accurate as in [12] for which the grid is uniform, for similar mesh refinement around the cylinder ($\Delta x \approx 0.01$).

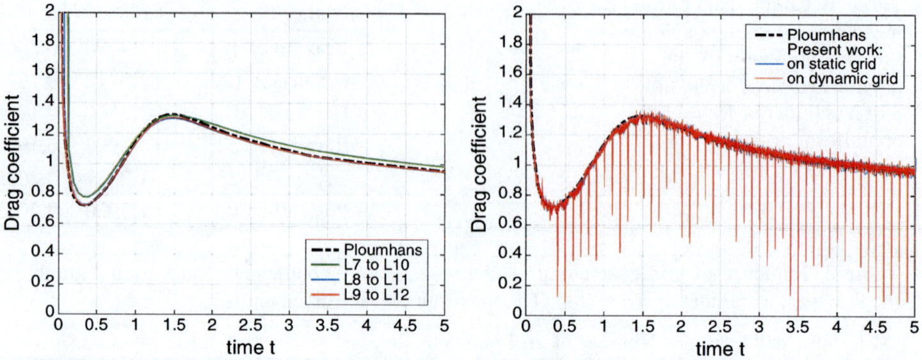

Fig. 7 Drag coefficients for the flow past a cylinder test at $Re = 550$ for both flow configurations. On the left: steady cylinder on different quadtree grids, on the right: impulsively started cylinder on static and dynamic $L_8 - L_{11}$ grids

For the impulsively started cylinder configuration, the simulations are run on static quadtree grids and on dynamic AMR grids. For the static quadtree grid, the most refined box is chosen such that the cylinder remains inside this box during the whole simulation. The dynamic adaptation of the grid is performed at each n iteration satisfying $n\Delta t = 0.1$. The time evolution of the drag coefficient C_D for both static and dynamic grids is also in good agreement with the results of Ploumhans [33]. The drag coefficient is oscillating because of the cylinder motion, but the amplitude of oscillations is decreasing in a consistent way when refining the grid. Moreover, a peak appears periodically when using dynamic grids, each time a mesh adaptation is performed. These spurious oscillations are due to the perturbation introduced between two consecutive time steps when the mesh is updated, i.e., all quantities are defined on the new mesh by interpolation of the quantities defined on the old mesh. More precisely, when four children merge into one parent (coarsening), an arithmetic average is performed to define data on the new coarse cell; when a parent is split into four children (refinement), the new four children all receive the data of the parent cell.

A snapshot of the vorticity at $t = 5$ is presented in Fig. 8. The adaptive grid is displayed in the background. The number of cells is considerably lower (about 4 times less) using this dynamic grid, which has a significant effect on the total calculation time. Using these AMR parameters, a simulation on the dynamic grid is about 65% less expensive than a simulation on the corresponding static quadtree grid. This loss of speed-up can be explained by the cost of each AMR process. As an indication, the computational cost of an AMR process is between 70 and 80% of the cost of a time iteration of the predictor–corrector scheme.

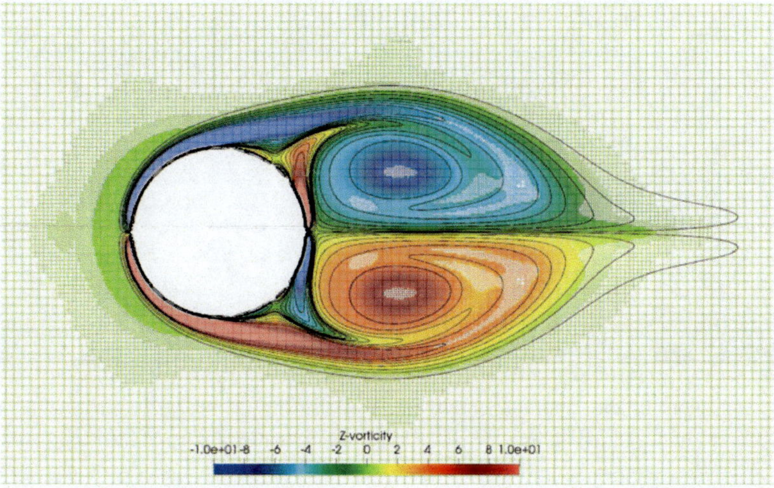

Fig. 8 Z-component of the vorticity ω at $t = 5$ for the impulsively started cylinder at $Re = 550$. The $L_9 - L_{12}$ grid is depicted in background. The contours are drawn for $|\omega(\mathbf{x})| = 0.1, 0.5, 1, 2, 3, 4, 5$

4.2 Sedimentation of a Cylinder

We are now focusing on the sedimentation of a cylinder, subject to gravity. A narrow cavity $\Omega = [0, 2] \times [0, 6]$ is filled by a fluid of density $\rho_f = 1$ and viscosity $\mu_f = 0.01$. Inside, a cylinder of radius $r = 1/8$ and center $x_c(t = 0) = (1, 4)$ is initially at rest. Its density is set to $\rho_s = 1.5$. The gravitational acceleration is $g = 980$.

Since the channel is narrow, the computational domain under consideration $\Omega = [-2, 4] \times [-0.1, 5.9]$ is mainly composed of penalized areas, and the grid is consequently coarsened as much as possible outside the cavity. No-slip boundary conditions are naturally imposed by penalization on the left, right, and bottom boundaries. Moreover, homogeneous Neumann conditions are prescribed at the top boundary.
Dynamic AMR grids are used. Next results, obtained on dynamic grids, are nearly identical to those obtained on uniform grids. The AMR process is performed at least every 0.005 s.

The time evolution of the velocity and position of the center of mass $\mathbf{x}_c(t)$ is described by the Newton–Euler equations. For simplicity, the velocity $\mathbf{u}_s(t)$ is defined as the translation velocity of the center of mass. Therefore, the $\mathbf{x}_c(t)$ and $\mathbf{u}_s(t)$ are determined by solving

$$\begin{aligned} \rho_s|\Omega_s|\frac{d\mathbf{u}_s}{dt} &= \mathbf{F}_{hydro} + (\rho_f - \rho_s)|\Omega_s|\mathbf{g} \\ \frac{d\mathbf{x}_c}{dt} &= \mathbf{u}_s, \end{aligned} \tag{25}$$

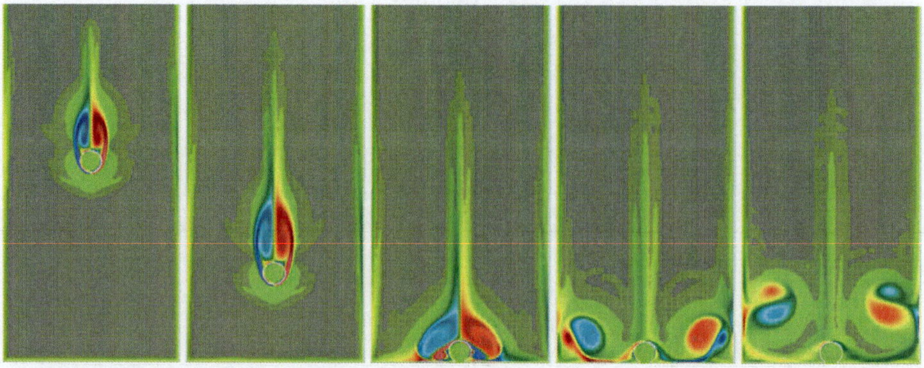

Fig. 9 Snapshots of the z-component of the vorticity during the sedimentation of a cylinder. From left to right, $t \in \{0.2, 0.3, 0.4, 0.5, 0.6\}$. The dynamic $L_8 - L_{10}$ AMR grid is depicted in background

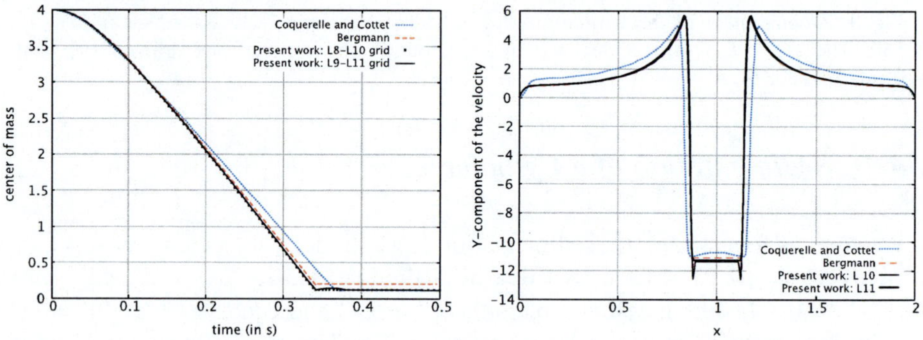

Fig. 10 Results obtained for the sedimentation of a cylinder test. Time evolution of the Y-component of the center of mass $\mathbf{x}_c$ (on the left) and vertical velocity profile along y-axis $y = y_c$ at time $t = 0.1$ (on the right). Comparison with the studies of Bergmann [12] and Coquerelle [34]

where $\mathbf{F}_{hydro}$ is the hydrodynamical force acting on the cylinder, $|\Omega_s|$ is the area of the cylinder, and $\mathbf{g} = (0, g)$ is the gravitational force. These ODEs are solved using the classical second-order Runge–Kutta scheme. No collision models are used in this solver. In this simulation, the CFL number is set to 0.1.

We compare the results obtained with the study of Bergmann [12] and Coquerelle [34]. Snapshots of the vorticity are presented in Fig. 9. The results presented in Fig. 10 are in good agreement with the results of the literature.

5 Conclusion

We have presented a versatile finite-volume method to simulate incompressible flows with rigid bodies. This immersed boundary approach is particularly dedicated to the consideration of arbitrary geometries that can move over time. The behavior of the bodies is imposed through a volume penalization term, and the interface is followed implicitly using a level-set function.

This work is based on finite-volume discretizations on quadtree grids, with compact stencils. For this purpose, the Bitpit library is an efficient tool for the generation of hierarchical adaptive grids, dedicated to massively parallel computations. The use of hierarchical meshes is particularly interesting because the adaptation of the mesh is algorithmically and computationally efficient. We use a criterion based on the Hessian of the velocity, which is a good indicator to estimate the regularity of the solution.

The validation of the model is performed on different test cases, including static or moving structures. The results obtained are in agreement with the previous studies, for both fixed and dynamic meshes. By adapting the mesh to the numerical solution, we can speed up greatly the computations if compared to fully uniform Cartesian methods, while ensuring an equivalent accuracy.

References

1. Johansen, J., Sørensen, N.N., Michelsen, J.A., Schreck, S.: Detached-eddy simulation of flow around the NREL Phase VI blade. Wind Energy Int. J. Progr. Appl. Wind Power Convers. Technol. **5**(2–3), 185–197 (2002)
2. Poullis, M.: Computational fluid dynamic analysis to prevent aortic root and valve clots during left ventricular assist device support. J. Extra-Corporeal Technol. **44**(4), 210 (2012)
3. Hu, H.H., Patankar, N.A., Zhu, M.Y.: Direct numerical simulations of fluid-solid systems using the arbitrary Lagrangian-Eulerian technique. J. Comput. Phys. **169**(2), 427–462 (2001)
4. Mittal, R., Iaccarino, G.: Immersed boundary methods. Annu. Rev. Fluid Mech. **37**, 239–261 (2005)
5. Glowinski, R., Pan, T. W., Hesla, T.I., Joseph, D.D., Periaux, J.: A fictitious domain approach to the direct numerical simulation of incompressible viscous flow past moving rigid bodies: application to particulate flow. J. Comput. Phys. **169**(2), 363–426 (2001)
6. Angot, P., Bruneau, C.H., Fabrie, P.: A penalization method to take into account obstacles in incompressible viscous flows. Numer. Math. **81**(4), 497–520 (1999)
7. Peskin, C.S.: Numerical analysis of blood flow in the heart. J. Comput. Phys. **25**(3), 220–252 (1977)
8. Peskin, C.S.: The fluid dynamics of heart valves: experimental, theoretical, and computational methods. Annu. Rev. Fluid Mech. **14**(1), 235–259 (1982)
9. Pilliod, J.E., Jr., Puckett, E.G.: Second-order accurate volume-of-fluid algorithms for tracking material interfaces. J. Comput. Phys. **199**(2), 465–502 (2004)
10. Shen, J., Yang, X.: A phase-field model and its numerical approximation for two-phase incompressible flows with different densities and viscosities. SIAM J. Sci. Comput. **32**(3), 1159–1179 (2010)
11. Osher, S., Sethian, J.A.: Fronts propagating with curvature-dependent speed: algorithms based on Hamilton-Jacobi formulations. J. Comput. Phys. **79**(1), 12–49 (1988)
12. Bergmann, M., Hovnanian, J., Iollo, A.: An accurate Cartesian method for incompressible flows with moving boundaries. Commun. Comput. Phys. **15**(5), 1266–1290 (2014)
13. Mittal, R., Dong, H., Bozkurttas, M., Najjar, F.M., Vargas, A., Von Loebbecke, A.: A versatile sharp interface immersed boundary method for incompressible flows with complex boundaries. J. Comput. Phys. **227**(10), 4825–4852 (2008)

14. Marella, S., Krishnan, Liu, H., Udaykumar, H.S.: Sharp interface Cartesian grid method I: an easily implemented technique for 3D moving boundary computations. J. Comput. Phys. **210**(1), 1–31 (2005)
15. Kadoch, B., Kolomenskiy, D., Angot, P., Schneider, K.: A volume penalization method for incompressible flows and scalar advection–diffusion with moving obstacles. J. Comput. Phys. **231**(12), 4365–4383 (2012)
16. Chorin, A.J.: Numerical solution of the Navier-Stokes equations. Math. Comput. **22**(104), 745–762 (1968)
17. Temam, R.: Sur l'approximation de la solution des équations de Navier-Stokes par la méthode des pas fractionnaires (II). Archive for Rational Mechanics and Analysis **33**(5), 377–385 (1969)
18. Goda, K.: A multistep technique with implicit difference schemes for calculating two- or three-dimensional cavity flows. J. Comput. Phys. **30**(1), 76–95 (1979)
19. Morton, G.M.: A Computer Oriented Geodetic Data Base and a New Technique in File Sequencing. International Business Machines Company, Armonk (1966)
20. Ferziger, J.H., Peric, M.: Computational Methods for Fluid Dynamics, vol. 3, pp. 196–200. Springer, Berlin (2002)
21. Patankar, S.: Numerical Heat Transfer and Fluid Flow. CRC Press, Boca Raton (2018)
22. Rhie, C.M., Chow, W.L.: Numerical study of the turbulent flow past an airfoil with trailing edge separation. AIAA J. **21**(11), 1525–1532 (1983)
23. Zang, Y., Street, R.L., Koseff, J.R.: A non-staggered grid, fractional step method for time-dependent incompressible Navier-Stokes equations in curvilinear coordinates. J. Comput. Phys. **114**(1), 18–33 (1994)
24. Coudière, Y., Vila, J.P., Villedieu, P.: Convergence rate of a finite volume scheme for a two dimensional convection-diffusion problem. ESAIM: Math. Model. Numer. Anal. **33**(3), 493–516 (1999)
25. Delcourte, S., Domelevo, K., Omnes, P.: Discrete duality finite volume method for second order elliptic problems. pp. 447–458. Hermes Science Publishing, Stanmore (2005)
26. Rusanov, V.V.: The calculation of the interaction of non-stationary shock waves with barriers. Zhurnal Vychislitel'noi Matematiki i Matematicheskoi Fiziki **1**(2), 267–279 (1961)
27. Noca, F.: On the evaluation of time-dependent fluid-dynamic forces on bluff bodies. Doctoral dissertation, California Institute of Technology (1997)
28. Braza, M., Chassaing, P., Minh, H.H.: Numerical study and physical analysis of the pressure and velocity fields in the near wake of a circular cylinder. J. Fluid Mech. **165**, 79–130 (1986)
29. Henderson, R.D.: Details of the drag curve near the onset of vortex shedding. Phys. Fluids **7**(9), 2102–2104 (1995)
30. He, J.W., Glowinski, R., Metcalfe, R., Nordlander, A., Periaux, J.: Active control and drag optimization for flow past a circular cylinder: I. Oscillatory cylinder rotation. J. Comput. Phys. **163**(1), 83–117 (2000)
31. Bergmann, M.: Optimisation aérodynamique par réduction de modèle POD et contrôle optimal: application au sillage laminaire d'un cylindre circulaire. Doctoral dissertation, Vandoeuvre-les-Nancy, INPL (2004)
32. Taymans, C.: Solving Incompressible Navier-Stokes Equations on Octree grids: towards Application to Wind Turbine Blade Modelling. Doctoral dissertation, Bordeaux (2018)
33. Ploumhans, P., Winckelmans, G.S.: Vortex methods for high-resolution simulations of viscous flow past bluff bodies of general geometry. J. Comput. Phys. **165**(2), 354–406 (2000)
34. Coquerelle, M., Cottet, G.H.: A vortex level set method for the two-way coupling of an incompressible fluid with colliding rigid bodies. J. Comput. Phys. **227**(21), 9121–9137 (2008)

Fluid–Structure Interaction Using Volume Penalization and Mass-Spring Models with Application to Flapping Bumblebee Flight

Hung Truong, Thomas Engels, Dmitry Kolomenskiy, and Kai Schneider

Abstract Wing flexibility plays an essential role in the aerodynamic performance of insects due to the considerable deformation of their wings during flight under the impact of inertial and aerodynamic forces. These forces come from the complex wing kinematics of insects. In this study, both wing structural dynamics and flapping wing motion are taken into account to investigate the effect of wing deformation on the aerodynamic efficiency of a bumblebee in tethered flight. A fluid–structure interaction solver, coupling a mass-spring model for the flexible wing with a pseudo-spectral code solving the incompressible Navier–Stokes equations, is implemented for this purpose. We first consider a tethered bumblebee flying in laminar flow with flexible wings. Compared to the rigid model, flexible wings generate smaller aerodynamic forces but require much less power. Finally, the bumblebee model is put into a turbulent flow to investigate its influence on the force production of flexible wings.

1 Introduction

In recent years, the effect of wing flexibility on aerodynamic performance of flapping wings has drawn attention of researchers, scientists, and engineers. Compared to conventional airplanes with fixed wings, flapping wings have several aerodynamic advantages with the ability to create lift even at high angles of attack

H. Truong (✉) · K. Schneider
Aix-Marseille Université, CNRS, Centrale Marseille, Marseille, France
e-mail: dinh-hung.truong@univ-amu.fr; kai.schneider@univ-amu.fr

T. Engels
LMD-IPSL, Ecole Normale Supérieure-PSL, Paris, France
e-mail: thomas.engels@ens.fr

D. Kolomenskiy
Global Scientific Information and Computing Center, Tokyo Institute of Technology, Tokyo, Japan
e-mail: dmitry@gsic.titech.ac.jp

R. Deiterding et al. (eds.), *Cartesian CFD Methods for Complex Applications*,
SEMA SIMAI Springer Series 3, https://doi.org/10.1007/978-3-030-61761-5_2

due to the delayed stall of the leading edge vortex (LEV) [1]. These extraordinary capabilities of hovering and maneuverability have made bio-inspired flapping wings a strong candidate for developing human-engineered micro-air-vehicles (MAVs) with possible applications in environmental monitoring, surveillance, and security. However, in previous studies, the wings were usually considered as rigid to simplify the problem. The sophisticated interaction between the anisotropic wing structures and the surrounding unsteady flow makes the analysis of flapping flexible wings challenging, but at the same time intriguing. In the last two decades, with the dramatic improvement of measuring equipment as well as computing power, many experimental and numerical studies have investigated the effect of wing flexibility and drawn contradictory conclusions. Mountcastle and Combes [2] showed that passive deformations enhance lift production in bumblebees by artificially stiffening their wings using a micro-splint. Campos et al. [3] and Fu et al. [4] used experimental methods and found that highly flexible wings show significant tip-root lag that weakened vortices and reduce the force production. Du and Sun [5] solved the Navier–Stokes equations coupled with measured wing deformation data and compared with the rigid counterparts. They obtained a 10% increase in lift caused by the camber deformation and a 5% reduction in required power.

In our previous work [6], we considered a flexible bumblebee wing rotating around a hinge point at the angle of attack equal to 45°. The stiffness of the wing was varied to get two cases: flexible and highly flexible. We found that the flexible wing produces less lift than the rigid wing, but it has a better lift-to-drag ratio. On the other hand, the highly flexible wing experienced a strong tip-root lag caused by twisting and behaves poorly in term of aerodynamic performance with a much smaller lift and lift-to-drag ratio. Although the study provided us some ideas about the influence of wing flexibility on the force generation, the revolving motion remains too simple to fully represent the complicated dynamics of a flapping wing. After a short transition period, the revolving wing attains its steady state and its dynamic deformation hardly plays a role in the force production. The wing kinematics of insects is in reality more intricate with many characteristic features such as flapping amplitude, wingbeat frequency, angle of attack, etc. These features have strong impact on the ability of generating force of the wings. Kang and Shyy [7] showed that the ratio between the flapping frequency and the first natural frequency of a flexible wing can yield advanced, symmetric, or delayed rotation modes, which in turn alter the resulting lift. Zhao et al. [8] conducted experiments of simple isotropic flapping wings with varied stiffness values at different angles of attack. They found that at low angles of attack (20° to 60°), flexible wings have essentially the same aerodynamic performance as rigid wings, but they outperformed their rigid counterparts at high angles of attack (up to 90°).

Consequently, in this work, we will investigate the aerodynamic efficiency of the flexible bumblebee wing model [6] within a tethered flight context where the wing motion is prescribed using real bumblebee wing kinematics measured by the work of Dudley and Ellington [9]. The resulting force and power are compared with those of a rigid flat wing computed by Engels et al. [10]. The comparison shows the effects of the wing deformation on the aerodynamic forces of a flapping flight insect.

The remainder of the manuscript presents in Sect. 2 the numerical methods used for solving the governing equations of the flexible wing, the fluid flow, and its coupling. The numerical setup and the bumblebee model are given in Sect. 3, and the results of the numerical simulations are discussed in Sect. 4. Finally, some conclusions are drawn in Sect. 5.

2 Numerical Methods and Governing Equations

Modeling insect flight is a delicate topic due to the fact that one needs to model both the mechanical behavior of the wings by a solid solver and the surrounding flow by a fluid solver. The two solvers must then be coupled to study their interaction. This section presents how these three aspects can be handled numerically.

2.1 Solid Solver Using Mass-Spring System

Insect wings are sophisticated structures consisting of membranes and veins. The wings get their nonlinear anisotropic properties from a truss framework composed of horizontal and vertical veins connected by membranes [11]. This along with their small wing lengths (from mm to cm) makes it extremely challenging to model the mechanical behavior of insect wings. In our work, a mass-spring system is employed to mimic the dynamics of the complicated membrane–vein network by taking the different mechanical properties between veins and membranes into account [6]. While veins can be considered as rods that resist mainly the torsion and bending deformation, a membrane is fabric-like and behaves like a piece of cloth that resists against the extension deformation.

The mass-spring system has been around since the end of the twentieth century, and it is well known for its computational efficiency and ability of handling large deformation [12]. The wing is discretized using mass points m_i $(i = 1, \ldots, n)$ connected by springs. Among the different types of springs, our model is built only on extension and bending springs. The dynamic behavior of the mass-spring system, at a given time t, is defined by the position $\mathbf{x}_i$ and the velocity $\mathbf{v}_i$ of the mass point i, and they are governed by the equations given below:

$$\begin{aligned} \mathbf{F}_i^{\text{int}} + \mathbf{F}_i^{\text{ext}} &= m_i \mathbf{a}_i \qquad \text{for } i = 1 \ldots n \\ \mathbf{v}_i(t = 0) &= \mathbf{v}_{0,i} \\ \mathbf{x}_i(t = 0) &= \mathbf{x}_{0,i}, \end{aligned} \tag{1}$$

where n is the number of mass points, $\mathbf{F}_i^{\text{int}}$ is the internal force and $\mathbf{F}_i^{\text{ext}}$ is the external force acting on the ith mass point, and m_i and $\mathbf{a}_i$ are mass and acceleration of the ith mass point, respectively.

The system (1) is then advanced numerically in time by applying a second-order backward differentiation scheme with variable time steps [13]:

$$\mathbf{q}_i^{n+1} - \frac{(1+\xi)^2}{1+2\xi}\mathbf{q}_i^n + \frac{\xi^2}{1+2\xi}\mathbf{q}_i^{n-1} = \frac{1+\xi}{1+2\xi}\Delta t^n \mathbf{f}(\mathbf{q}_i^{n+1}), \tag{2}$$

where $\mathbf{q} = \left[\mathbf{x}_i,\ \mathbf{v}_i\right]^{\mathrm{T}}$ is the phase vector containing positions and velocities of all mass points, $\mathbf{f}(\mathbf{q}) = \left[\mathbf{v}_i,\ m_i^{-1}(\mathbf{F}_i^{\mathrm{int}} + \mathbf{F}_i^{\mathrm{ext}})\right]^{\mathrm{T}}$ is the right-hand side function, and $\xi = \Delta t^n / \Delta t^{n-1}$ is the ratio between the current time step Δt^n and the previous one Δt^{n-1}. The phase vector of the system at the current time step $\mathbf{q}^{n+1}$ is found by solving Eq. (2) using the Newton–Raphson method. All details of this solver are explained in [6].

2.2 Fluid Solver and Volume Penalization Method

Due to their small sizes and elevated flapping frequencies, insect flight is normally categorized in the Reynolds number regime between $O(10^1)$ and $O(10^4)$. For example, for hawkmoth, we have $Re = 6000$, bumblebee $Re = 2000$, fruit fly $Re = 100$, or thrips $Re = 10$ [14, 15]. The flow can be considered as incompressible and governed by

$$\partial_t \mathbf{u} + \boldsymbol{\omega} \times \mathbf{u} = -\nabla\Pi + \nu\nabla^2\mathbf{u} - \underbrace{\frac{\chi}{C_\eta}(\mathbf{u} - \mathbf{u}_s)}_{\text{penalization term}} - \underbrace{\frac{1}{C_{\mathrm{sp}}}\nabla \times \frac{(\chi_{sp}\boldsymbol{\omega})}{\nabla^2}}_{\text{sponge term}} \tag{3}$$

$$\nabla \cdot \mathbf{u} = 0 \tag{4}$$

$$\mathbf{u}(\mathbf{x}, t=0) = \mathbf{u}_0(\mathbf{x}) \qquad \mathbf{x} \in \Omega,\ t > 0. \tag{5}$$

The above equations (3–5) are called the penalized Navier–Stokes equations [16], where $\mathbf{u}$ is the fluid velocity, $\boldsymbol{\omega} = \nabla \times \mathbf{u}$ is the vorticity, $\Pi = p + \frac{1}{2}\mathbf{u} \cdot \mathbf{u}$ is the total pressure, and ν is the kinematic viscosity. In addition to all the terms found in the classical incompressible Navier–Stokes equations, there appear two more terms that are called the sponge and the penalization terms. The former is added to remove the periodicity of the Fourier discretization that affects the upstream inflow. The penalization term is used to impose the no-slip boundary conditions on the fluid–solid interface in [17]. All geometrical information of the solid is encoded in the mask function χ given by

$$\chi(\delta) = \begin{cases} 1 & \delta \le d - h \\ \frac{1}{2}\left(1 + \cos\pi\frac{(\delta - d + h)}{2h}\right) & d - h < \delta < d + h \\ 0 & \delta \ge d + h, \end{cases} \tag{6}$$

where δ is the signed distance field of the bumblebee "skeleton" and d represents here the distance from the "skeleton" to the outer surface, i.e., the fluid–solid interface. The "skeleton" of the bumblebee is a curvilinear centerline along which we sweep an elliptical section of variable size to draw the insect's body, legs, and antennae [18]. However, to avoid the force oscillation when dealing with moving solid body, a smoothing layer with a thickness $2h$ is added right at the fluid–solid interface to prevent the discontinuity of the mask function [19].

For solving the fluid equations (3–5), a Fourier pseudo-spectral discretization with semi-implicit time stepping is employed, implemented in the FLUSI[1] code [17]. The general idea consists of representing quantities q (velocity, pressure, and vorticity) as truncated Fourier series,

$$q(\mathbf{x},t) = \sum_{k_x=0}^{N_x-1} \sum_{k_y=0}^{N_y-1} \sum_{k_z=0}^{N_z-1} \hat{q}(\mathbf{k},t) \exp(i\mathbf{k}\cdot\mathbf{x}), \tag{7}$$

where $\mathbf{k} = [k_x, k_y, k_z]^{\mathrm{T}}$ is the wavevector, $i = \sqrt{-1}$, and $\widehat{q}$ are the discrete complex Fourier coefficients of q. The Fourier coefficients can be computed with the fast Fourier transform (FFT) using the P3DFFT library. The main motivation of using a Fourier discretization is the simplicity of inverting a diagonal Laplace operator and the high numerical precision reflected in the absence of numerical diffusion and dissipation in the discretization. The gradient of a scalar can, for instance, be obtained by multiplying with the wavevector and the complex unit, $\widehat{\nabla q} = i\mathbf{k}\hat{q}$. The Laplace operator becomes a simple multiplication by $-|\mathbf{k}|^2$, and it is thus diagonal in Fourier space. For further details, we refer the reader to the reference article on the FLUSI solver [17].

2.3 *Fluid–Structure Interaction*

For time-stepping, the coupled fluid–solid system is advanced by employing a semi-implicit staggered scheme, as proposed in [18]. On the one hand, we advance the fluid by using the Adam–Bashforth second-order (AB2) scheme with exact integration of the viscous term. On the other hand, the backward differentiation formula of second order (BDF2) is used for the time discretization of the solid solver. The two modules are then coupled by the algorithm presented in the flowchart shown in Fig. 1. For the range of Reynolds numbers (75–4000), Dickinson et al. [20–22] showed that pressure forces dominate the shear viscous forces. Hence, for calculating the solid deformation, the viscous fluid tension is considered negligible compared to the static pressure. Moreover, the scheme is called a weak

[1] FLUSI: freely available for noncommercial use from GitHub (https://github.com/pseudospectators/FLUSI).

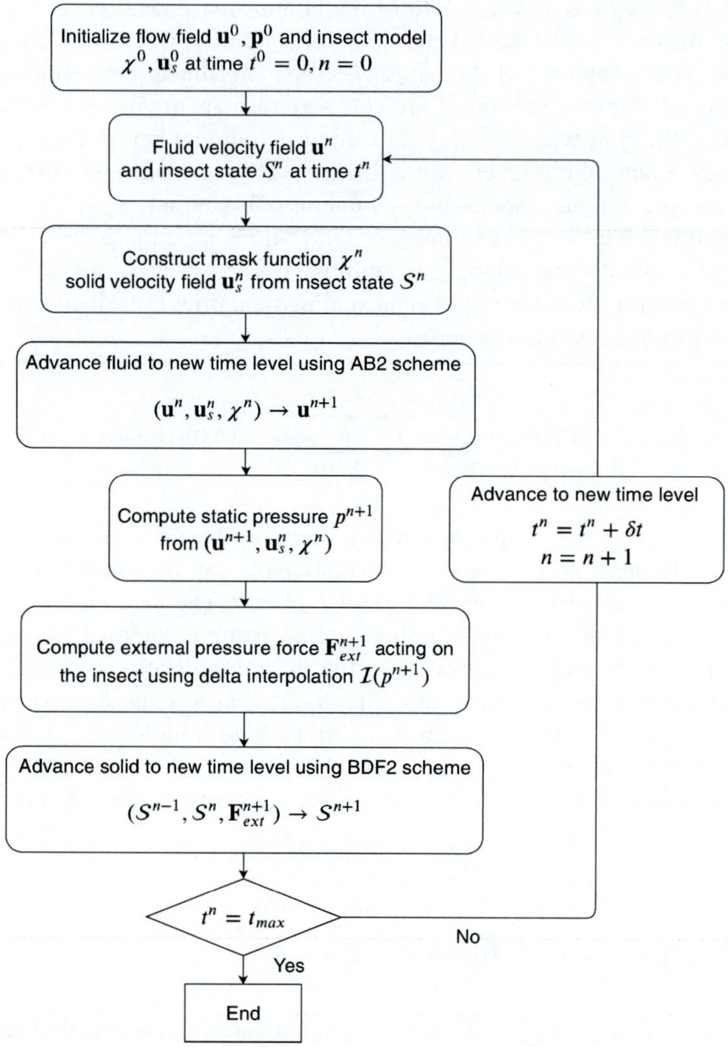

Fig. 1 Flowchart: semi-implicit staggered scheme of the time advancement for the fluid–structure interaction problem

coupling method since the static pressure is computed from the previous state of the solid model. This makes the system conditionally stable only if the structure is heavy enough with respect to the fluid density. However, the scheme is efficient because the fluid and the solid need to be advanced only one time at the current time level. Full details of the fluid–structure interaction (FSI) framework as well as detailed validation of the results can be found in our previous work [6].

3 Numerical Setup and Bumblebee Model

To study the influence of wing flexibility on the aerodynamic forces, we compare the flexible wings with rigid ones using the same numerical setup as in the previous work [10].

3.1 Flow Configuration

The computational domain, shown in Fig. 2, is $6R \times 4R \times 4R$ large, where R is the bumblebee wing length, discretized by $1152 \times 768 \times 768$ grid points. The bumblebee is tethered (both translational and rotational motions of the body are constrained) at $\mathbf{x}_{\text{cntr}} = (2R, 2R, 2R)^T$ and exposed to a head wind with a mean flow accounting for the insect's forward velocity $\mathbf{u}_\infty = (1.246Rf, 0, 0)^T$, where f is the wingbeat frequency. Due to the periodicity inherent to the spectral method, a thin vorticity sponge outlet, covering the last 4 grid points in x-direction, is used to minimize the upstream influence of the computational domain. The sponge

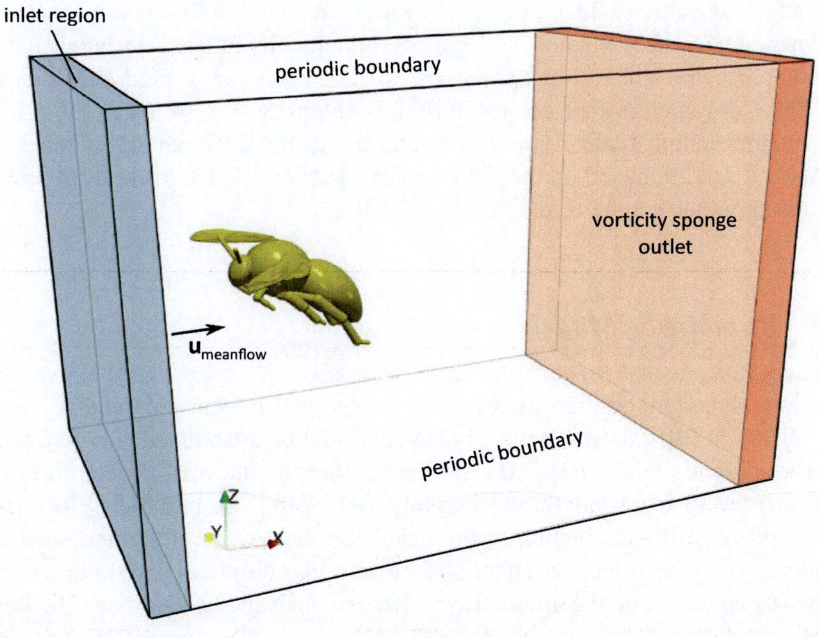

Fig. 2 Illustration of the computational domain of size $6R \times 4R \times 4R$ used in all simulations. A bumblebee model with flexible wings is tethered at $\mathbf{x}_{cntr} = (2R, 2R, 2R)^T$ in a flow with the mean flow velocity $\mathbf{u}_\infty = (1.246Rf, 0, 0)^T$. The blue inlet region is used to impose a precomputed HIT velocity field that is constantly advected downstream by the mean flow. Another vorticity sponge region is placed at the outlet to damp out vortices. Periodic boundary conditions are set for the four other sides of the domain

penalization parameter C_{sp} is usually set to a value larger than the permeability C_η, normally $C_{sp} = 10^{-1}$. By construction, the sponge term is divergence-free to avoid the influence on the pressure field, which in turn would be modified even in regions far away from the sponge due to its nonlocality. A detailed discussion on the influence of the vorticity sponge can be found in [17].

In nature, insects do not always fly in a calm, quiescent environment. Instead, they face, most of the time, many kinds of aerial perturbations such as gusty wind, vortices, or turbulent flow generated by surrounding obstacles. Taking this into account, both laminar and turbulent flows are investigated here to study the role of wing flexibility under these two circumstances. For the laminar case, in the entire computational domain, a mean flow $\mathbf{u}_\infty$ is imposed by simply setting the zeroth Fourier mode of the velocity $\mathbf{u}$ [18]. On the other hand, information on turbulent flow conditions, which are experienced by flying insects in nature, remains an open question with limited data [23]. However, for indoor wind tunnel experiments, isotropic or near-isotropic turbulence generated by a grid has been used as inflow condition to study the impact of turbulence on insect flight performance. Consequently, homogeneous isotropic turbulence (HIT) is chosen as turbulent inflow in our present work in order to compare with the results obtained for rigid wings in [10, 15]. For this purpose, in the inlet region containing the first 48 grid points along the axial direction, a precomputed HIT velocity field $\mathbf{u}'$ is added to the mean flow as velocity fluctuations $\mathbf{u}_{in} = \mathbf{u}_\infty + \mathbf{u}'$. The HIT field is then transported downstream by the mean flow and evolves dynamically like grid turbulence. In order to compare with the results from [10], we use here a HIT field characterized by the same parameters that are the turbulent intensity $Tu = u'_{RMS}/u_\infty = 0.33$, the integral length scale $\Lambda = 0.77R$, and the turbulent Reynolds number $Re_\lambda = u_{RMS}\lambda/\nu = 129$, based on the Taylor microscale $\lambda = 0.18$. More technical details on this approach can be found in [10, 15, 18].

3.2 Bumblebee Model

The bumblebee model here is the same as the one used in [15] and derived from case BB01 in [9], except for the wings which will be introduced later in Sect. 3.3.

The animal's body mass, M, is 175 mg, the gravitational acceleration is $g = 9.81\,\mathrm{m/s^2}$, and wing length, R, amounts to 15 mm. The bumblebee is composed of linked rigid bodies including the head, the thorax, the abdomen, all legs, the proboscis, and the antennae. These parts are circular elliptical or cylindrical sections joined by spheres, and the bilateral symmetry of the insect is assumed. The Reynolds number is $Re = U_{tip}c_m/\nu_{air} = 2685$, where $U_{tip} = 2\Phi Rf = 9.15\,\mathrm{m/s}$ is the mean wingtip velocity, $c_m = 4.6\,\mathrm{mm}$ is the mean chord length, $\nu_{air} = 1.568 \cdot 10^{-5}\mathrm{m^2/s}$ is the kinematic viscosity of air, $f = 152\,\mathrm{Hz}$ ($T = 1/f = 6.6\,\mathrm{ms}$) is the wingbeat frequency (T is duration), and $\phi = 115°$ is the wingbeat amplitude. The wingbeat kinematics is prescribed based on the work of Dudley and Ellington [9].

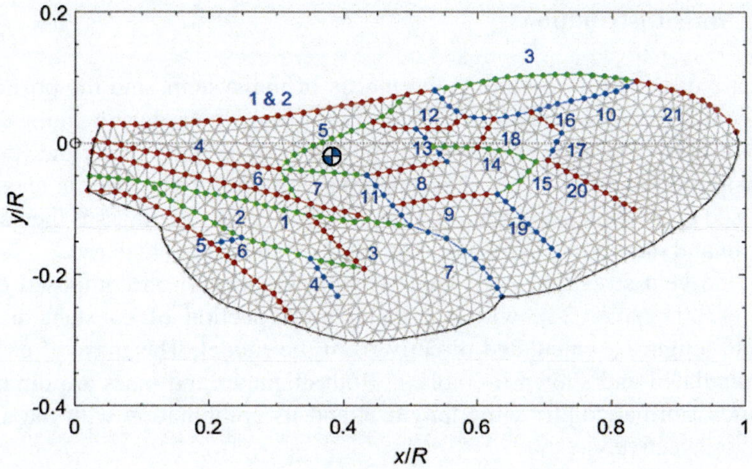

Fig. 3 Illustration of the mass-spring model that is meshed based on measured data of real bumblebee wings [24]. The blue and white marker represents the mass center. Color codes (red, green, and blue) are used for identifying the veins, and the membranes are represented by gray circles

3.3 *Flexible Wing Model*

The two flexible wings of the insect are modeled using the mass-spring system as detailed in [6]. In the following, we describe the venation pattern, the mass distribution, and the flexural rigidity of the veins.

3.3.1 Venation Pattern

The venation architecture is claimed to be responsible for the anisotropy of the wing, and it plays a crucial role on the wing dynamics during flight. Consequently, the functional approach is used to take into account the venation pattern in our model. The wing contour and the vein network are adapted from [24] and encoded into the mass-spring system. The wing is then discretized by a triangular mesh with 1065 mass points, as shown in Fig. 3, using SALOME,[2] an open-source integration platform for mesh generation. A mesh convergence study comparing between two wings, discretized by 465 and 1065 mass points, was performed in [6] for the revolving motion. Looking at the aerodynamic forces generated, the coarse-mesh wing showed no major difference with respect to the fine-mesh wing. However, for the flapping motion, the pressure field is expected to be more unstable and a fine-mesh wing is needed for the pressure interpolation in this case.

[2]https://www.salome-platform.org/.

3.3.2 Mass Distribution

The mass distribution represents the inertia of the system, and the position of the mass center has a connection with the wing dynamics during flight. The mass distribution is calculated based on the measured wing mass data from [24] and the vein pattern. For our numerical simulations, the total wing mass is chosen as the same used by Kolomenskiy et al. [24], $m_w = 0.791$ mg. The mass is then distributed into vein and membrane parts based on their geometry and material.

For the vein structure, each vein is considered as a rod composed of cuticle, $\rho_c = 1300\,\text{kg/m}^3$ [24], with a circular cross section of constant diameter d_v [24] and length l_v, calculated directly from the model. The mass of each vein is then calculated and shown in Table 1. Both diameter and mass are dimensionless quantities, normalized by wing length R and its combination with the air density $\rho_{air} R^3$, respectively.

For the mass distribution of the membrane, the same optimization method as in [6] is applied, where the objective function is the difference between the mass center of the wing measured in the experiment [24] and the one calculated from the mass-

Table 1 Dimensionless vein diameter d_v data (adapted from [24]) and the corresponding dimensionless mass m_v data

Forewing			Hindwing		
#	Nominal diameter	Nominal mass	#	Nominal diameter	Nominal mass
1	0.0070	0.0209	1	0.0065	0.0180
2	0.0074	0.0237	2	0.0043	0.0071
3	0.0055	0.0076	3	0.0046	0.0024
4	0.0070	0.0063	4	0.0011	0.0001
5	0.0040	0.0031	5	0.0038	0.0043
6	0.0048	0.0094	6	0.0037	0.0005
7	0.0040	0.0019	7	0.0020	0.0012
8	0.0038	0.0009			
9	0.0041	0.0023			
10	0.0048	0.0064			
11	0.0045	0.0017			
12	0.0038	0.0018			
13	0.0042	0.0010			
14	0.0038	0.0020			
15	0.0034	0.0008			
16	0.0032	0.0005			
17	0.0032	0.0004			
18	0.0044	0.0009			
19	0.0015	0.0001			
20	0.0018	0.0001			
21	0.0020	0.0009			

spring model. For a mass point m_i belonging to the membrane at position $[x_i, y_i]$, we obtain

$$m_i = 9.14 \cdot 10^{-5} - 3.48 \cdot 10^{-5} x_i + 2.48 \cdot 10^{-4} y_i. \quad (8)$$

Differences, between two mass centers, of $3.85 \cdot 10^{-3}[R]$ in the x-direction and $0.93 \cdot 10^{-3}[R]$ in the y-direction are obtained. These are negligible compared to the reference wing length R.

3.3.3 Flexural Rigidity of Veins

Because the bending stiffness of the membrane is neglected, the flexural rigidity of the wing comes solely from the flexural rigidity EI of veins, which is calculated based on their material and geometry. While the estimation of their second moments of inertia I is straightforward using the diameter data from Table 1, determining the Young's modulus is not trivial. In our present work, the veins are considered to be made of cuticle, which is reported to have a Young's modulus in the range of 1 kPa to 50 MPa [25]. The wing needs to be flexible enough to reveal the influence of wing flexibility to the aerodynamic performance of insects, but it cannot be too flexible to show unrealistic mechanical behaviors. For the purpose of our study, the value $E = 700$ kPa is chosen.

4 Results and Discussion

The forces generated by the bumblebee model with flexible wings as well as the required aerodynamic power will be presented in this section. Furthermore, they will be compared with the results obtained in [10], where the same bumblebee with rigid wings was considered. This allows us to have some insight into the wing flexibility influence on the insect aerodynamic performance.

4.1 Tethered Flight in Laminar Flow

The vertical and horizontal forces produced by the flapping motion of the flexible wings are shown by red curves in Fig. 4a and b, while blue curves are those generated by rigid wings. The forces are normalized by $F = \rho_{air} R^4 f^2$. Here, the sideways force is small and not presented, since the animal is modeled with the assumption of symmetry. The simulation is computed for 4 strokes with 28,776 time steps using 32 processors on Intel Xeon Gold 6142 (Sky Lake) 2.6 GHz and consumed 8128 CPU hours. For each cycle, the cycle-averaged values are calculated and presented in Table 2. While the wing flexibility has minor effect on the average

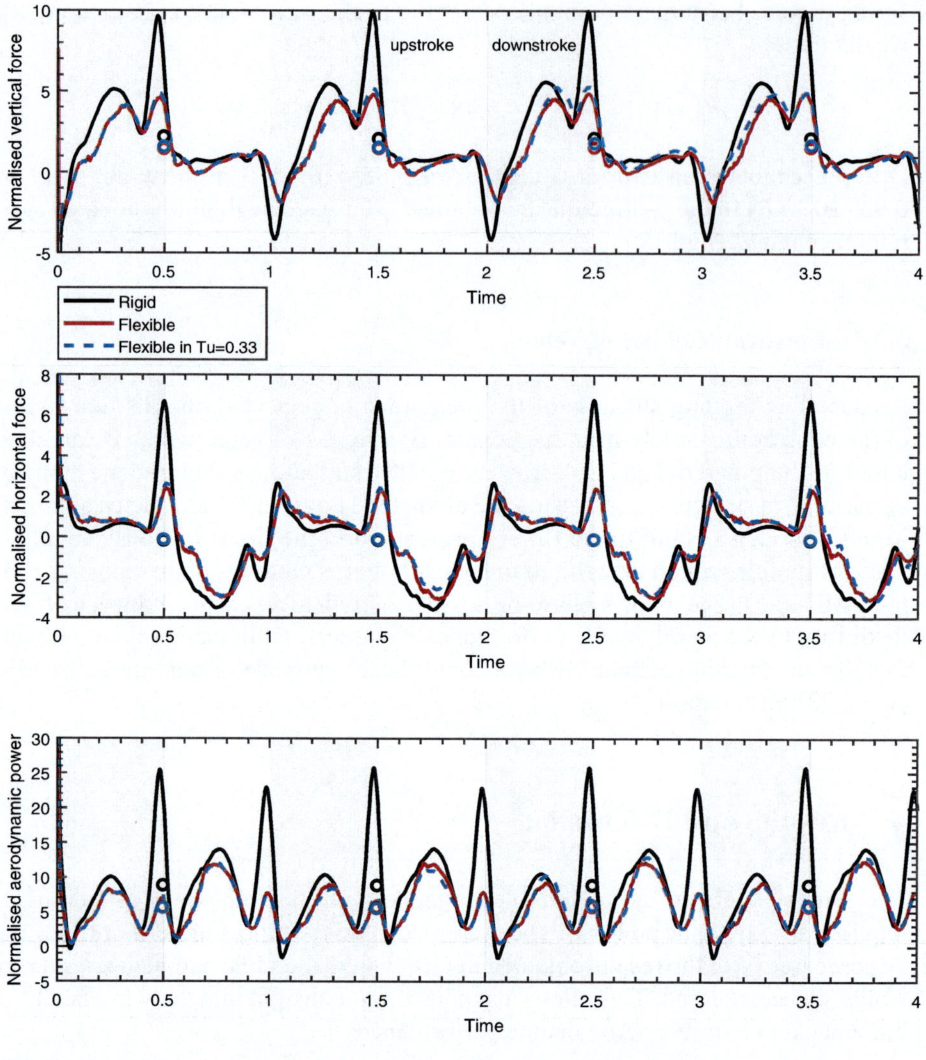

Fig. 4 Normalized vertical force, horizontal force, and aerodynamic power generated by a bumblebee with rigid wings (black) [15] and flexible wings in laminar flow (red) and turbulent flow (blue). Circles represent the cycle-averaged value of forces and power

thrust with a decline of 11%, it accounts for a 28% drop of the average lift. These losses can be explained as a result from the decrease of the effective angle of attack caused by wing deformation. The shape adaptation of the wing during the flapping motion alters the instantaneous angle of attack, which is claimed to play a significant role in the force generation [4]. However, these negative impacts do not necessarily mean that the rigid wings outperform aerodynamically their flexible counterparts. Although the flexible wings generate smaller forces, they consume much less energy, with almost 36% required aerodynamic power is reduced. The

Table 2 Cycle-averaged forces and power in the laminar case

	Thrust		Lift		Aerodynamic power	
Flow	Rigid	Flexible	Rigid	Flexible	Rigid	Flexible
Laminar	0.17	0.15	2.09	1.51	8.84	5.67

cycle-averaged lift-to-power ratio of flexible wing is 0.026, 35% larger than that of rigid wing which is 0.019.

Nevertheless, regarding the time evolution of the forces during one wingbeat, the instant surges of the forces at the ends of upstroke and downstroke, observed in the rigid case, are significantly weakened. The sudden rotation of the rigid wings at the middle and the end of strokes is the reason for these large force peaks [26]. This effect has now little impact due to the fact that the wing inertia is now taken into account. The inertial force makes the wing deform and streamline its shape to the airflow. This shape adaptation helps to mitigate the large pressure jump between upper and lower surfaces, especially at the trailing edge [7], and provides a smoother flight [27, 28]. This finding has more advantages in term of stabilizing generated forces, rather than lift-enhancement effect.

4.2 *Tethered Flight in Turbulent Flow*

We then study the influence of isotropic turbulence on the aerodynamic performance of a bumblebee by putting it into a turbulent flow. The simulation is computed for 4 strokes with 29,000 time steps using 32 processors on Intel Xeon Gold 6142 (Sky Lake) 2.6 GHz and consumed 9000 CPU hours. Figure 5 presents the flow structure of the bumblebee flying in a turbulent flow visualized by the normalized vorticity isosurfaces at two levels, $\|\boldsymbol{\omega}\| = 15$ and $\|\boldsymbol{\omega}\| = 100$. The aerodynamic forces and the corresponding power for this turbulent flow condition are shown in Fig. 4. The results demonstrate insignificant differences between turbulent and laminar flow conditions. The aerodynamic forces generated by the bumblebee are almost identical to those derived during unperturbed, laminar inflow, with the same required energetic cost. For $Re > 100$, the aerodynamic forces are mainly produced by the differential dynamics pressure across the wing [26]. Figure 6 shows the normalized pressure distribution on top and bottom wing surfaces of the two cases just before the stroke reversal $t = 0.45\,\mathrm{T}$. The effect of turbulence can hardly be seen here, which explains the negligible change of aerodynamic forces. The outcome here is consistent with the one observed in the rigid case in [10].

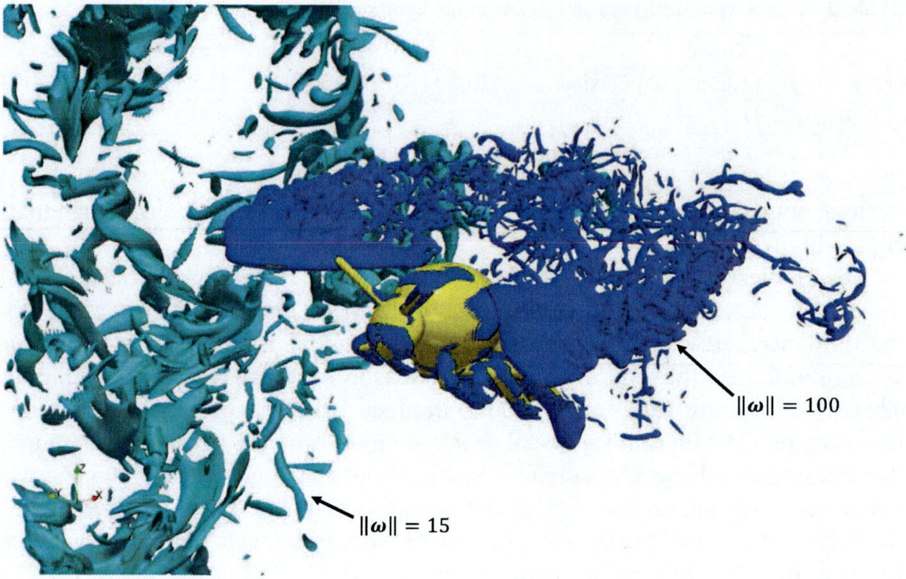

s

Fig. 5 Visualization of flow generated by a tethered flapping bumblebee with flexible wings in turbulence with turbulent intensity $Tu = 0.33$ showing normalized absolute vorticity isosurfaces at two levels, $||\omega|| = 15$ (light blue) and $||\omega|| = 100$ (dark blue). The flow fields are plotted at time $t = 0.45/T$, and the weaker vortices are only shown in the region $3.7R \leq y \leq 4R$

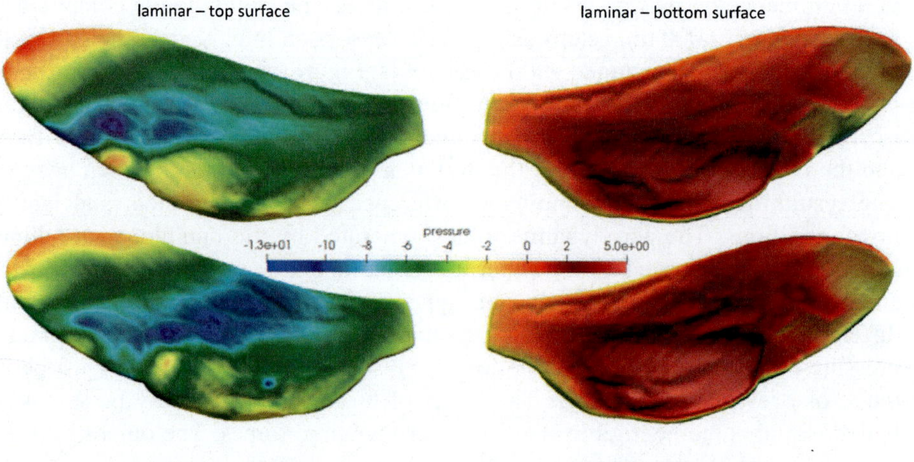

Fig. 6 Normalized pressure distribution on the wing for top and bottom surfaces, plotted at time $t = 0.45/T$ just before the stroke reversal, for the laminar (top) and turbulent cases (bottom)

5 Conclusions and Perspectives

Following our previous work on revolving flexible wings [6], the impact of wing flexibility was now studied in the context of tethered flight using flapping wing kinematics measured in experiments by Dudley and Ellington [9]. High-resolution numerical simulations on massively parallel machines were carried out to solve the fluid–structure interaction problem between the fluid solver FLUSI and the solid solver based on a mass-spring system. Both laminar and turbulent inflows were considered to investigate diverse flight conditions of insects. The preliminary results obtained in this work allow us to have some understanding about the role of wing flexibility in flapping flight.

In laminar flow, the aerodynamic forces and the required power have been calculated and compared with the ones obtained for rigid wings. We found that wing flexibility hardly contributed to lift or thrust enhancement. However, the significant reduction of the required power suggested that wing flexibility plays an important role in saving flight energetic cost. Moreover, the wing inertia also helped to damp out the fluctuation of the aerodynamic force and thus helped the insect to stabilize during flight.

In turbulent flow, although the ability of shape adaptation of flexible wings makes them more sensitive to fluctuation of the flow structure than their rigid counterparts, the impact of turbulence is still negligible under the considered flight conditions. Nevertheless, due to costly computational time, the statistical property of the turbulent flow is not considered because only one simulation is done to obtain the results for the turbulent case. Moreover, due to the expensive computational cost, especially in the turbulent case, the mesh convergence study was not performed in this paper, and we refer readers to [6].

Despite of these findings, we have to keep in mind that the wing kinematics has an essential effect on the aerodynamic performance of wings, and we have considered only one set of wing motion in this study. In perspective, these limitations can be overcome by examining other species with different wing kinematics or including flight control in our model. This is planned for our work in the future where we will study Calliphora with its wing kinematics measured from experiments.

Finally, although the wing flexibility was calculated based on the geometrical property of the veins, the estimation of veins' Young's modulus remains somewhat limited due to the vast range of known cuticle properties. This can be improved by using mathematical optimization for determining the right elastic properties of the wing model. To this end, the equilibrium state of the wing model under external static force as a function of wing stiffness will be calculated and compared with data measured from experiments done by our team.

Acknowledgments The authors gratefully acknowledge financial support from the Agence Nationale de la Recherche (ANR Grant No. 15-CE40-0019) and Deutsche Forschungsgemeinschaft (DFG Grant No. SE 824/26-1), project AIFIT. The authors were granted access to the HPC resources of IDRIS under the Allocation No. 2018-91664 attributed by GENCI (Grand Équipement National de Calcul Intensif). For this work, Centre de Calcul Intensif d'Aix-Marseille is acknowledged for granting access to its high performance computing resources financed by the project Equip@Meso (No. ANR-10-EQPX- 29-01). The authors thankfully acknowledge financial support granted by the ministères des Affaires étrangères et du développement international (MAEDI) et de l'Education nationale et l'enseignement supérieur, de la recherche et de l'innovation (MENESRI), and the Deutscher Akademischer Austauschdienst (DAAD) within the French–German Procope project FIFIT.

D.K. gratefully acknowledges financial support from the JSPS KAKENHI Grant No. JP18K13693.

References

1. Ellington, C.P., van den Berg, C., Willmott, A.P., Thomas, A.L.R.: Leading-edge vortices in insect flight. Nature **384**, 626–630 (1996)
2. Mountcastle, A.M., Combes, S.A.: Wing flexibility enhances load-lifting capacity in bumblebees. Proc. R. Soc. B. **280**(1759) (2016)
3. Campos, D., Ukeiley, L., Bernal, L.: Flow around flapping flexible flat plate wings. In: 50th AIAA Aerospace Sciences Meeting including the New Horizons Forum and Aerospace Exposition (2012). https://doi.org/10.2514/6.2012--1211
4. Fu, J., Liu, X., Shyy, W., Qiu, H.: Effects of flexibility and aspect ratio on the aerodynamic performance of flapping wings. Bioinspir. Biomim. **13**(3), 036001 (2018)
5. Du, G., Sun, M.: Effects of wing deformation on aerodynamic forces in hovering hoverflies. J. Exp. Biol. **213**(13), 2273–2283 (2010)
6. Truong, H., Engels, T., Kolomenskiy, D., Schneider, K.: A mass-spring fluid-structure interaction solver: application to flexible revolving wings. Comput. Fluid. **200**, 104426 (2020)
7. Kang, C.-K., Shyy, W.: Scaling law and enhancement of lift generation of an insect-size hovering flexible wing. J. R. Soc. Inter. **10**(85), 20130361 (2013)
8. Zhao, L., Huang, Q., Deng, X., Sane, S.P.: Aerodynamic effects of flexibility in flapping wings. J. R. Soc. Inter. **7**(44), 485–497 (2010)
9. Dudley, R., Ellington, C.P.: Mechanics of forward flight in bumblebees I. kinematics and morphology. J. Exp. Biol. **148**, 19–52 (1990)
10. Engels, T., Kolomenskiy, D., Schneider, K, Lehmann, F.O., Sesterhenn, J.: Bumblebee flight in heavy turbulence. Phys. Rev. Lett. **116**, 028103 (2016)
11. Shyy, W., Aono, H., Kang, C., Liu, H.: An Introduction to Flapping Wing Aerodynamics. Cambridge University Press, New York (2013)
12. Nealen, A., Muller, M., Keiser, R., Boxerman, E., Carlson, M.: Physically based deformable models in computer graphics. Comput. Graph. Forum **25**, 809–836 (2006)
13. Berger, J.: A second order backward difference method with variable steps for a parabolic problem. BIT Numer. Math. **38**, 644–662 (1998)
14. Shyy, W., Kang, C.K., Chirarattananon, P., Ravi, S., Liu, H.: Aerodynamics, sensing and control of insect-scale flapping-wing flight. Proc. Math. Phys. Eng. Sci. **472**(2186), 20150712 (2016)
15. Engels, T., Kolomenskiy, D., Schneider, K., Farge, M., Lehmann, F.O., Sesterhenn, J.: Impact of turbulence on flying insects in tethered and free flight: high-resolution numerical experiments. Phys. Rev. Fluid. **4**, 013103 (2019)
16. Angot, P., Bruneau, C., Fabrie, P.: A penalization method to take into account obstacles in incompressible viscous flows. Numer. Math. **81**, 497–520 (1999)
17. Engels, T., Kolomenskiy, D., Schneider, K., Sesterhenn, J.: Flusi: a novel parallel simulation tool for flapping insect flight using a Fourier method with volume penalization. SIAM J. Sci. Comp. **38**, S03–S24 (2016)

18. Engels, T.: Numerical modeling of fluid-structure interaction in bioinspired propulsion. Ph.D. Thesis at Aix-Marseille Université and TU Berlin (2015). https://hal.archives-ouvertes.fr/tel-01298968
19. Kolomenskiy, D., Schneider, K.: A Fourier spectral method for the Navier-Stokes equations with volume penalization for moving solid obstacles. J. Comput. Phys. **228**, 5687–5709 (2009)
20. Dickinson, M.H., Lehmann, F.-O., Sane, S.P.: Wing rotation and the aerodynamic basis of insect flight. Science **284**(5422), 1954–1960 (1999)
21. Dickinson, M.H., Götz, K.: Unsteady aerodynamic performance of model wings at low Reynolds numbers. J. Exp. Biol. **174**, 45–64 (1993)
22. Roccia, B.A., Preidikman, S., Massa, J.C., Mook, D.T.: Modified unsteady vortex-lattice method to study flapping wings in hover flight. AIAA J. **51**(11), 2628–2642 (2013)
23. Crall, J.D., Chang, J.J., Oppenheimer, R.L., Combes, S.A.: Foraging in an unsteady world: bumblebee flight performance in field-realistic turbulence. Inter. Focus **7**, 20160086 (2017)
24. Kolomenskiy, D., Ravi, S., Xu, R., Ueyama, K., Jakobi, T., Engels, T., Nakata, T., Sesterhenn, J., Schneider, K., Onishi, R., Liu, H.: The dynamics of passive feathering rotation in hovering flight of bumblebees. J. Fluids. Struct. **91**, 102628 (2019)
25. Vincent, J.F.V., Wegst, U.G.K.: Design and mechanical properties of insect cuticle. Arthropod Struct. Dev. **33**, 187–199 (2004)
26. Sane, S.P.: The aerodynamics of insect flight. J. Exp. Biol. **206**, 4191–4208 (2003)
27. Ifju, P.G., Jenkins, A.D., Ettingers, S., Lian, Y., Shyy, W.: Flexible-wing-based micro air vehicles. In: 40th AIAA Aerospace Sciences Meeting and Exhibit. Reno, January 14–17 (2002)
28. Ifju, P.G., Peter, G., Stanford, B., Sytsma, M.: Analysis of a flexible wing micro air vehicle. In: Proceedings of 25th AIAA Aerodynamic Measurement Technology and Ground Testing Conference. San Francisco, June 5–8 (2006)

No-Slip and Free-Slip Divergence-Free Wavelets for the Simulation of Incompressible Viscous Flows

Souleymane Kadri Harouna and Valérie Perrier

Abstract This work concerns divergence-free wavelet-based methods for the numerical resolution of Navier–Stokes equations. It generalizes to higher dimension the approach of Kadri-Harouna and Perrier (Multiscale Model. Simul. 13:399–422; 2015) that reformulates the projection method using the Helmholtz–Hodge decomposition in wavelet domain. The solution is searched in a finite dimensional free-slip divergence-free wavelet space, with time-dependent wavelet coefficients. We prove and verify the convergence of a first-order time numerical scheme for the Helmholtz–Hodge-based projection method. Numerical simulations on the 3D lid-driven cavity flow show the accuracy and efficiency of the method.

1 Introduction

The numerical resolution of the time-dependent Navier–Stokes equations for an incompressible viscous fluid still remains a complex problem. The direct numerical simulation where all the flows of eddies are simulated is very costly in terms of computational time and memory storage resources. One reason of such a difficulty is that, commonly, the velocity field and the pressure are coupled in the numerical discretization due to the incompressibility constraint of the velocity field [16, 34]. Moreover, the numerical methods for the resolution of mixed problems (Stokes problem) require an inf–sup condition to be satisfied for the velocity and pressure discretization spaces [16, 35]. In practice, it is difficult to manage both discretizations to obtain this inf–sup condition. Thus, the conditional number of the arising system is very high, which increases the numerical cost of the global method.

S. K. Harouna (✉)
Laboratoire de Mathématiques Image et Applications, La Rochelle, France
e-mail: souleymane.kadri_harouna@univ-lr.fr

V. Perrier
Univ. Grenoble Alpes, CNRS, Grenoble INP, LJK, Grenoble, France
e-mail: Valerie.Perrier@univ-grenoble-alpes.fr

R. Deiterding et al. (eds.), *Cartesian CFD Methods for Complex Applications*, SEMA SIMAI Springer Series 3, https://doi.org/10.1007/978-3-030-61761-5_3

By their property of data sparse representation, wavelet-based methods have been introduced in the numerical resolution of the Stokes problem to get robust and effective numerical schemes, with lower data storage. Principally, wavelet bases linked by differentiation and integration allow to stabilize the spatial discretizations and get the inf–sup condition [2, 31]; another interest lies in the ability to provide adaptive strategies to reduce the algorithm complexity [3, 6, 7, 13–15, 30, 31, 35].

The projection method initialized by Chorin and Temam [4, 33] is an approach that avoids the difficulties of the Stokes problem: it is based on a time-splitting method that uncouples the computation of the velocity field from the pressure. An intermediate velocity is computed, and then this predicted velocity is projected onto the space of divergence-free vector fields. The simplicity of the method lies in the fact that the prediction and the correction steps are elliptic problems, namely Poisson equations. However, the projection method introduces an additional numerical splitting error, which must be at worst of the same order as the time discretization error. In addition, the corrected velocity field does not satisfy the desired boundary condition, and the projection step imposes an artificial boundary condition on the pressure [38].

A reformulation of the projection method was proposed in [22] consisting in a change of variable like in the Gauge method [26, 37]. The main idea of [22] is to replace the classical correction step by a Helmholtz–Hodge decomposition of the intermediate predicted velocity field, using a divergence-free wavelet basis. This allows to encode boundary conditions into the divergence-free wavelet basis and avoids the use of non-physical boundary conditions for the pressure. The numerical complexity of the divergence-free wavelet-based Helmholtz–Hodge decomposition does not exceed the numerical complexity of the resolution of a Poisson equation, see [20].

The achievements of the present paper are to propose an extension of [22] to the three-dimensional case, to provide a convergence result and 3D numerical simulations. First, we describe a recent construction of divergence-free wavelets, more easy to handle, satisfying no-slip and free-slip boundary conditions on the hypercube in dimension 3, borrowed from [23]. Then, following the approach of [22] for the 2D case, we use these divergence-free wavelets to design a numerical method for the resolution of Navier–Stokes equations in 3D. As done for the 2D (linear) Stokes equations in [22], we study here the convergence of the method for the (nonlinear) 3D Navier–Stokes equations, using a first-order time-discretization scheme. Then, a standard error analysis allows us to prove the convergence of the numerical scheme, under a CFL type condition. Numerical experiments conducted on benchmark flows confirm these theoretical results.

This paper is organized as follows. Section 2 summarizes the principle of the construction of divergence-free wavelet bases. We recall in Sect. 2.1 the construction of the one-dimensional biorthogonal multiresolution analyses linked by differentiation and integration that we used to construct divergence-free wavelets satisfying physical boundary conditions in Sect. 2.2. Section 3 presents the numerical method, and we investigate the convergence of the time-discretization scheme in the energy norm. Section 4 describes our fully discrete scheme and presents numerical simulation results on benchmark flow, particularly the 3D lid-driven cavity flow.

2 Free-Slip Divergence-Free Wavelet Bases on $[0, 1]^d$

Since the seminal works of Lemarié-Rieusset and collaborators [19, 25], several constructions of divergence-free (and curl-free) wavelet bases on $[0, 1]^d$ have been provided in the literature [8, 21, 31, 32, 35, 36]. The free-slip boundary condition case was treated by several methods in [21, 31, 32]. In this section, we follow the construction principle of [21, 23], which we extend to general dimension. We begin with our construction of biorthogonal multiresolution analyses of $L^2(0, 1)$ linked by differentiation and integration.

2.1 *Multiresolution Analyses Linked by Differentiation and Integration*

Divergence-free wavelet constructions on the hypercube require the use of multiresolution analyses on the interval linked by differentiation and integration [19, 21, 31]. Specifically, we would like to have at hand two biorthogonal multiresolution analyses of $L^2(0, 1)$, denoted by $(V_j^1, \tilde{V}_j^1)$ and $(V_j^0, \tilde{V}_j^0)$ satisfying

$$\frac{d}{dx} V_j^1 = V_j^0. \tag{1}$$

The construction of the biorthogonal spaces $\tilde{V}_j^1$ and $\tilde{V}_j^0$ is of the utmost importance. A suitable choice would provide the commutation of the multi-scale projectors with the derivation operator. Such a choice was suggested in [19, 21]:

$$\tilde{V}_j^0 = \left\{ \int_0^x f(t)dt \; : \; f \in \tilde{V}_j^1 \right\} \cap H_0^1(0, 1) \Rightarrow \frac{d}{dx} \tilde{V}_j^0 \subset V_j^0. \tag{2}$$

In this case, we have

$$\frac{d}{dx} \circ \mathcal{P}_j^1 = \mathcal{P}_j^0 \circ \frac{d}{dx}, \tag{3}$$

where $\mathcal{P}_j^1$ is biorthogonal projector onto V_j^1 and $\mathcal{P}_j^0$ the biorthogonal projector onto V_j^0. A significant property of such a construction is that the spaces $\tilde{V}_j^0$ provide a multiresolution analysis for $H_0^1(0, 1)$ (and not for $L^2(0, 1)$). However, the edge wavelets of [21] do not satisfy a diagonal differentiation relation. A remedy to this was proposed in [31, 32], where the commutation property of the multi-scale projectors and the derivation operator is lost. Another construction proposed in [23] consists in setting

$$\frac{d}{dx}\tilde{V}_j^0 = \tilde{V}_j^d, \tag{4}$$

where $\tilde{V}_j^d$ is the biorthogonal space of the space satisfying homogeneous Dirichlet boundary conditions $V_j^d = V_j^1 \cap H_0^1(0,1)$ and $\tilde{V}_j^d \not\subset H_0^1(0,1)$. This choice preserves the commutation property and leads to a diagonal differentiation relation for the wavelet bases even for edge wavelets.

The constructions of [19, 21] are based on multiresolution analyses of $L^2(0,1)$ reproducing polynomials at boundaries [5, 27, 28]. Each space is spanned by a scaling functions basis

$$V_j^1 = \text{span}\{\varphi_{j,k}^1 \ ; \ 0 \le k \le N_j - 1\} \text{ and } \tilde{V}_j^1 = \text{span}\{\tilde{\varphi}_{j,k}^1 \ ; \ 0 \le k \le N_j - 1\},$$

and

$$V_j^0 = \text{span}\{\varphi_{j,k}^0 \ ; \ 0 \le k \le N_j - 2\} \text{ and } \tilde{V}_j^0 = \text{span}\{\tilde{\varphi}_{j,k}^0 \ ; \ 0 \le k \le N_j - 2\},$$

with dimension parameter $N_j = 2^j + c$, for $c \in \mathbb{Z}$. For $\varepsilon = 0, 1$, the scaling functions $\varphi_{j,k}^\varepsilon$ can be written as $\varphi_{j,k}^\varepsilon = 2^{j/2}\varphi^\varepsilon(2^j x - k)$ inside the interval $[0,1]$, where φ^ε is a compactly supported scaling function on $\mathbb{R}$, but this is no more true near the boundaries 0 and 1 (idem for $\tilde{\varphi}_{j,k}^\varepsilon$). In practice, the scale index j must be greater than some index j_{min}, to avoid boundary effects [28]. The biorthogonality between bases writes

$$\langle \varphi_{j,k}^\varepsilon, \tilde{\varphi}_{j,k'}^\varepsilon \rangle = \delta_{k,k'}.$$

The advantage of using multiresolution analyses reproducing polynomials at boundaries is that homogeneous boundary conditions can be easily incorporated. It suffices to remove the scaling functions that do not satisfy the desired condition at edges 0 and 1, prior to biorthogonalization [27, 28]. This writes

$$V_j^d = V_j^1 \cap H_0^1(0,1) = \text{span}\{\varphi_{j,k}^1 \ ; \ 1 \le k \le N_j - 2\}, \tag{5}$$

and

$$V_j^{dd} = V_j^1 \cap H_0^2(0,1) = \text{span}\{\varphi_{j,k}^1 \ ; \ 2 \le k \le N_j - 3\}. \tag{6}$$

Again, for the construction of the biorthogonal spaces $\tilde{V}_j^d$ and $\tilde{V}_j^{dd}$, we proceed similarly by removing edge scaling functions, to ensure the equality of dimensions and this leads to $(V_j^d, \tilde{V}_j^d)$ and $(V_j^{dd}, \tilde{V}_j^{dd})$, biorthogonal multiresolution analyses of $H_0^1(0,1)$ and $H_0^2(0,1)$, respectively [27]. Notice that the spaces $\tilde{V}_j^d$ and $\tilde{V}_j^{dd}$ provide multiresolution analyses of $L^2(0,1)$, see [23, 27].

Following the standard constructions [5, 17], the wavelets in the biorthogonal multiresolution analysis $(V_j^1, \tilde{V}_j^1)$ are defined as the bases of the biorthogonal spaces:

$$W_j^1 = V_{j+1}^1 \cap (\tilde{V}_j^1)^\perp \text{ and } \tilde{W}_j^1 = \tilde{V}_{j+1}^1 \cap (V_j^1)^\perp.$$

These spaces are finite dimensional [5]:

$$W_j^1 = \text{span}\{\psi_{j,k}^1 \; ; \; 0 \le k \le 2^j - 1\} \text{ and } \tilde{W}_j^1 = \text{span}\{\tilde{\psi}_{j,k}^1 \; ; \; 0 \le k \le 2^j - 1\},$$

with

$$\langle \psi_{j,k}^1, \tilde{\psi}_{j',k'}^1 \rangle = \delta_{j,j'}\delta_{k,k'} \text{ and } \langle \psi_{j,k}^1, \tilde{\varphi}_{j',k'}^1 \rangle = 0.$$

Then, working in $H_0^1(0,1)$, the first possibility is to construct the wavelet spaces as in [27, 28]:

$$W_j^d = V_{j+1}^d \cap (\tilde{V}_j^d)^\perp \text{ and } \tilde{W}_j^d = \tilde{V}_{j+1}^d \cap (V_j^d)^\perp.$$

This possibility was at the origin of the construction [21]. But unfortunately, the diagonal relation with the derivation is lost for edge wavelets. To overcome this difficulty, another approach was proposed by Kadri-Harouna and Perrier [23] and consists in first using a standard construction for the wavelet bases of $W_j^0 = V_{j+1}^0 \cap (\tilde{V}_j^0)^\perp$ and $\tilde{W}_j^0 = \tilde{V}_{j+1}^0 \cap (V_j^0)^\perp$[5, 17, 27, 28], denoted by $\{\psi_{j,k}^0\}_{j \ge j_{min}}$ and $\{\tilde{\psi}_{j,k}^0\}_{j \ge j_{min}}$. In the second step, the wavelets of W_j^d and $\tilde{W}_j^d$ are defined by

$$\psi_{j,k}^d = 2^j \int_0^x \psi_{j,k}^0 \qquad \text{and} \qquad 2^{-j}(\tilde{\psi}_{j,k}^0)' = -\tilde{\psi}_{j,k}^d. \tag{7}$$

As remarked before, this definition is different from that of [19, 21], where the wavelets $\psi_{j,k}^d$ and $\tilde{\psi}_{j,k}^d$ were defined first and second one set: $\psi_{j,k}^0 = 2^{-j}(\psi_{j,k}^d)'$ and $\tilde{\psi}_{j,k}^d = -2^j \int_0^x \tilde{\psi}_{j,k}^d$.

Now, due to the vanishing moments of $\psi_{j,k}^0$, it is easy to see that definition (7) implies $\psi_{j,k}^d \in H_0^1(0,1)$, and for any $j > j_{min}$,

$$V_j^d = V_{j_{min}}^d \oplus W_{j_{min}}^d \oplus \cdots \oplus W_{j-1}^d. \tag{8}$$

Decomposition (8) is stable in $H_0^1(0,1)$, i.e., the system $\{\psi_{j,k}^d\}_{j \ge j_{min}} \cup \{\varphi_{j_{min},k}^d\}$ is a Riesz basis for $H_0^1(0,1)$, see [23].

For each basis, a fast wavelet transform algorithm exists with a linear complexity and the approximation order of each multiresolution analysis space is linked to the number of vanishing moments of its biorthogonal wavelets [5, 21, 27, 28].

Precisely, if φ^1 allows reproduction of polynomials up to degree $r-1$ in V_j^1, then the biorthogonal wavelet $\tilde{\psi}^1$ has r vanishing moments:

$$\int_{-\infty}^{+\infty} x^\ell \tilde{\psi}^1(x)dx = 0 \text{ for } 0 \le \ell \le r-1.$$

Then, due to the differentiation and integration relations (1) and (4), the wavelet $\tilde{\psi}^0$ has $(r-1)$ vanishing moments and ψ^0 has $(\tilde{r}+1)$ vanishing moments.

2.2 Free-Slip and No-Slip Divergence-Free Wavelet Construction

In this section, we recall the construction of free-slip divergence-free wavelets on the hypercube $[0,1]^3$ as proposed in [20, 23]. Then, we outline the construction of no-slip divergence-free wavelets. The next section will sketch the general form of no-slip divergence-free wavelets to higher space dimension $d > 3$.

The divergence-free function space with free-slip boundary conditions is

$$\mathcal{H}_{div}(\Omega) = \{\boldsymbol{u} \in (L^2(\Omega))^3 : \nabla \cdot \boldsymbol{u} = 0 \text{ and } \boldsymbol{u} \cdot \boldsymbol{n}|_{\partial\Omega} = 0\}, \tag{9}$$

where $\Omega = [0,1]^3$ and $\boldsymbol{n}$ denotes the unit outward normal to boundary $\partial\Omega$. Then, following [20], there are three kinds of divergence-free scaling functions in $\mathcal{H}_{div}(\Omega)$ defined by

$$\Phi_{j,k}^{div,1} := \textbf{curl}\begin{bmatrix} 0 \\ 0 \\ \varphi_{j,k_1}^d \otimes \varphi_{j,k_2}^d \otimes \varphi_{j,k_3}^0 \end{bmatrix} = \begin{bmatrix} \varphi_{j,k_1}^d \otimes (\varphi_{j,k_2}^d)' \otimes \varphi_{j,k_3}^0 \\ -(\varphi_{j,k_1}^d)' \otimes \varphi_{j,k_2}^d \otimes \varphi_{j,k_3}^0 \\ 0 \end{bmatrix}, \tag{10}$$

$$\Phi_{j,k}^{div,2} := \textbf{curl}\begin{bmatrix} \varphi_{j,k_1}^0 \otimes \varphi_{j,k_2}^d \otimes \varphi_{j,k_3}^d \\ 0 \\ 0 \end{bmatrix} = \begin{bmatrix} 0 \\ \varphi_{j,k_1}^0 \otimes \varphi_{j,k_2}^d \otimes (\varphi_{j,k_3}^d)' \\ -\varphi_{j,k_1}^0 \otimes (\varphi_{j,k_2}^d)' \otimes \varphi_{j,k_3}^d \end{bmatrix}, \tag{11}$$

$$\Phi_{j,k}^{div,3} := \textbf{curl}\begin{bmatrix} 0 \\ \varphi_{j,k_1}^d \otimes \varphi_{j,k_2}^0 \otimes \varphi_{j,k_3}^d \\ 0 \end{bmatrix} = \begin{bmatrix} -\varphi_{j,k_1}^d \otimes \varphi_{j,k_2}^0 \otimes (\varphi_{j,k_3}^d)' \\ 0 \\ (\varphi_{j,k_1}^d)' \otimes \varphi_{j,k_2}^0 \otimes \varphi_{j,k_3}^d \end{bmatrix}. \tag{12}$$

By construction, these scaling functions are in $\mathcal{H}_{div}(\Omega)$ and the space $\mathbf{V}_j^{div}$ that they spanned is included in the multiresolution analysis of $(L^2(\Omega))^3$ given by

$$\mathbf{V}_j^{div} \subset \mathbf{V}_j^d = \left(V_j^d \otimes V_j^0 \otimes V_j^0\right) \times \left(V_j^0 \otimes V_j^d \otimes V_j^0\right) \times \left(V_j^0 \otimes V_j^0 \otimes V_j^d\right). \tag{13}$$

Moreover, let $\mathbf{P}_j$ be the biorthogonal multiscale projector onto $\mathbf{V}_j^d$:

$$\mathbf{P}_j = \left(\mathcal{P}_j^d \otimes \mathcal{P}_j^0 \otimes \mathcal{P}_j^0\right) \times \left(\mathcal{P}_j^0 \otimes \mathcal{P}_j^d \otimes \mathcal{P}_j^0\right) \times \left(\mathcal{P}_j^0 \otimes \mathcal{P}_j^0 \otimes \mathcal{P}_j^d\right),$$

and

$$\mathbf{P}_j^0 = \mathcal{P}_j^0 \otimes \mathcal{P}_j^0 \otimes \mathcal{P}_j^0,$$

the biorthogonal multiscale projector onto $\mathbf{V}_j^0 = V_j^0 \otimes V_j^0 \otimes V_j^0$. Due to the commutation property (3), we have

$$\forall\, \boldsymbol{u} \in \mathcal{H}_{div}(\Omega),\ \nabla \cdot \mathbf{P}_j(\boldsymbol{u}) = \mathbf{P}_j^0(\nabla \cdot \boldsymbol{u}) = 0.$$

In terms of spaces, this rewrites $\mathbf{P}_j(\mathcal{H}_{div}(\Omega)) = \mathbf{V}_j \cap \mathcal{H}_{div}(\Omega)$. Therefore, the spaces $\mathbf{V}_j^{div} = \mathbf{V}_j \cap \mathcal{H}_{div}(\Omega)$ provide a multiresolution analysis of $\mathcal{H}_{div}(\Omega)$, and a basis of $\mathbf{V}_j^{div}$ is generated by choosing two of the scaling function generators listed above [23]. Besides, from each scaling function generator, one can construct 7 types of anisotropic divergence-free wavelets [8, 23, 25]. We denote by $\{\Psi_{\boldsymbol{j},\boldsymbol{k}}^{div,1}, \Psi_{\boldsymbol{j},\boldsymbol{k}}^{div,2}\}$ the wavelets corresponding to $\{\Phi_{j,\boldsymbol{k}}^{div,1}, \Phi_{j,\boldsymbol{k}}^{div,2}\}$ and $\mathbf{W}_j^{div}$ the space they span:

$$\mathbf{W}_{\boldsymbol{j}}^{div} = span\left\{\Psi_{\boldsymbol{j},\boldsymbol{k}}^{div,1}, \Psi_{\boldsymbol{j},\boldsymbol{k}}^{div,2}\right\}_{j_{min} \le |\boldsymbol{j}| \le j-1}, \tag{14}$$

where the index $\boldsymbol{k} = (k_1, k_2, k_3)$ varies depending on the corresponding function: if one coordinate of $\boldsymbol{k}$ corresponds to a wavelet $\psi_{j,k}^d$ or $\psi_{j,k}^0$, we take $1 \le k \le 2^j$; if it corresponds to a scaling function $\varphi_{j,k}^d$, we take $1 \le k \le N_j - 2$; and if it corresponds to a scaling function $\varphi_{j,k}^0$, we take $0 \le k \le N_j - 2$. In the sequel, we will use this convention for all the summation over the index $\boldsymbol{k}$.

According to (14), the anisotropic multiscale decomposition of $\mathbf{V}_j^{div}$ corresponds to

$$\mathbf{V}_j^{div} = \mathbf{V}_{j_{min}}^{div} \bigoplus_{j_{min} \le |\boldsymbol{j}| \le j-1} \mathbf{W}_{\boldsymbol{j}}^{div}, \tag{15}$$

and since the spaces $\mathbf{V}_j^{div}$ provide a multiresolution analysis of $\mathcal{H}_{div}(\Omega)$, decomposition (15) is stable for this space. Specifically, we have the following proposition [23]:

Proposition 1 *For every function $\boldsymbol{u} \in \mathcal{H}_{div}(\Omega)$, there are coefficients $c_{\boldsymbol{k}}^{div,\epsilon}$ and $d_{\boldsymbol{j},\boldsymbol{k}}^{div,\epsilon}$ such that*

$$\boldsymbol{u} = \sum_{\epsilon=1}^{2} \left(\sum_{\boldsymbol{k}} c_{\boldsymbol{k}}^{div,\epsilon} \Phi_{j_{min},\boldsymbol{k}}^{div,\epsilon} + \sum_{|\boldsymbol{j}| \geq j_{min}} \sum_{\boldsymbol{k}} d_{\boldsymbol{j},\boldsymbol{k}}^{div,\epsilon} \Psi_{\boldsymbol{j},\boldsymbol{k}}^{div,\epsilon} \right), \tag{16}$$

and for two positive constants C_1 and C_2 independent of $\boldsymbol{u}$, we have

$$C_1 \|\boldsymbol{u}\|_{L^2} \leq \left(\sum_{\epsilon=1}^{2} \sum_{\boldsymbol{k}} |c_{\boldsymbol{k}}^{div,\epsilon}|^2 + \sum_{\epsilon=1}^{2} \sum_{|\boldsymbol{j}| \geq j_{min}} \sum_{\boldsymbol{k}} |d_{\boldsymbol{j},\boldsymbol{k}}^{div,\epsilon}|^2 \right)^{1/2} \leq C_2 \|\boldsymbol{u}\|_{L^2}. \tag{17}$$

Proposition 1 is immediate using Lemma 2 borrowed from [19, 25] and reported in the Appendix section. Precisely, the constructed divergence-free wavelets and their duals satisfy the hypothesis of this lemma, see [23].

The approximation error provided by the divergence-free wavelet basis is linked to the approximation order of the spaces V_j^1. If the spaces V_j^1 contain polynomials up to degree $r-1$, then V_j^0 contain polynomials up to degree $r-2$ and for all $\boldsymbol{u} \in (H^s(\Omega))^3 \cap \mathcal{H}_{div}(\Omega)$ with $0 \leq s \leq r-1$, and the following Jackson type estimation holds:

$$\|\boldsymbol{u} - \mathbb{P}_j^{div}(\boldsymbol{u})\|_{L^2} \leq C 2^{-js} \|\boldsymbol{u}\|_{H^s}, \tag{18}$$

where $\mathbb{P}_j^{div}$ is the biorthogonal multiscale projector onto $\mathbf{V}_j^{div}$:

$$\mathbb{P}_j^{div}(\boldsymbol{u}) = \sum_{\epsilon=1}^{2} \left(\sum_{\boldsymbol{k}} c_{\boldsymbol{k}}^{div,\epsilon} \Phi_{j_{min},\boldsymbol{k}}^{div,\epsilon} + \sum_{j_{min} \leq |\boldsymbol{j}| \leq j-1} \sum_{\boldsymbol{k}} d_{\boldsymbol{j},\boldsymbol{k}}^{div,\epsilon} \Psi_{\boldsymbol{j},\boldsymbol{k}}^{div,\epsilon} \right). \tag{19}$$

For homogeneous Dirichlet boundary conditions, the divergence-free function space is slightly different and will be denoted by

$$\mathcal{H}_{div,0}(\Omega) = \left\{ \boldsymbol{u} \in (H_0^1(\Omega))^3 \,:\, \nabla \cdot \boldsymbol{u} = 0 \right\} = (H_0^1(\Omega))^3 \cap \mathcal{H}_{div}(\Omega).$$

The space $\mathcal{H}_{div,0}(\Omega)$ is a closed subspace of $(H_0^1(\Omega))^3$, then we have the following decomposition:

$$(H_0^1(\Omega))^3 = \mathcal{H}_{div,0}(\Omega) \oplus \mathcal{H}_{div,0}(\Omega)^{\perp},$$

which is an orthogonal decomposition with respect to the inner product of $(H_0^1(\Omega))^3$: $(\boldsymbol{u}, \boldsymbol{v})_{(H_0^1(\Omega))^2} = (\nabla \boldsymbol{u}, \nabla \boldsymbol{v})_{(L^2(\Omega))^3}$, see [16] for details.

Since $\mathcal{H}_{div,0}(\Omega) \subset \mathcal{H}_{div}(\Omega)$, a multiresolution analysis of $\mathcal{H}_{div,0}(\Omega)$ is then provided by the spaces:

$$\mathbf{V}_j^{div,0} = \mathbf{V}_j^{dd} \cap \mathcal{H}_{div,0}(\Omega),$$

where

$$\mathbf{V}_j^{dd} = \left(V_j^{dd} \otimes V_j^{00} \otimes V_j^{00}\right) \times \left(V_j^{00} \otimes V_j^{dd} \otimes V_j^{00}\right) \times \left(V_j^{00} \otimes V_j^{00} \otimes V_j^{dd}\right), \tag{20}$$

and $V_j^{00} = V_j^0 \cap H_0^1(0,1)$. A scaling function basis of $\mathbf{V}_j^{div,0}$ is then given by

$$\Phi_{j,k}^{div,0,1} := \mathbf{curl} \begin{bmatrix} 0 \\ 0 \\ \varphi_{j,k_1}^{dd} \otimes \varphi_{j,k_2}^{dd} \otimes \varphi_{j,k_3}^{00} \end{bmatrix} = \begin{bmatrix} \varphi_{j,k_1}^{dd} \otimes (\varphi_{j,k_2}^{dd})' \otimes \varphi_{j,k_3}^{00} \\ -(\varphi_{j,k_1}^{dd})' \otimes \varphi_{j,k_2}^{dd} \otimes \varphi_{j,k_3}^{00} \\ 0 \end{bmatrix}, \tag{21}$$

$$\Phi_{j,k}^{div,0,2} := \mathbf{curl} \begin{bmatrix} \varphi_{j,k_1}^{00} \otimes \varphi_{j,k_2}^{dd} \otimes \varphi_{j,k_3}^{dd} \\ 0 \\ 0 \end{bmatrix} = \begin{bmatrix} 0 \\ \varphi_{j,k_1}^{00} \otimes \varphi_{j,k_2}^{dd} \otimes (\varphi_{j,k_3}^{dd})' \\ -\varphi_{j,k_1}^{00} \otimes (\varphi_{j,k_2}^{dd})' \otimes \varphi_{j,k_3}^{dd} \end{bmatrix}, \tag{22}$$

$$\Phi_{j,k}^{div,0,3} := \mathbf{curl} \begin{bmatrix} 0 \\ \varphi_{j,k_1}^{dd} \otimes \varphi_{j,k_2}^{00} \otimes \varphi_{j,k_3}^{dd} \\ 0 \end{bmatrix} = \begin{bmatrix} -\varphi_{j,k_1}^{dd} \otimes \varphi_{j,k_2}^{00} \otimes (\varphi_{j,k_3}^{dd})' \\ 0 \\ (\varphi_{j,k_1}^{dd})' \otimes \varphi_{j,k_2}^{00} \otimes \varphi_{j,k_3}^{dd} \end{bmatrix}, \tag{23}$$

where $\{\varphi_{j,k}^{00}\}_{1 \le k \le N_j - 3} = \{\varphi_{j,k}^{0}\}_{1 \le k \le N_j - 3}$ is the scaling function basis of V_j^{00} [28].

Similarly, to construct no-slip divergence-free wavelets associated with $\mathbf{V}_j^{div,0}$, it suffices to replace the wavelets of W_j^d by those of $W_j^{dd} = V_{j+1}^{dd} \cap (\tilde{V}_j^{dd})^{\perp}$ in the formula that defines the functions $\{\Psi_{j,k}^{div,\epsilon}\}_{\epsilon=1,2,3}$, see [21, 32]. Let now $\mathbb{P}_j^{div,0}$ be the L^2-orthogonal projector from $(H_0^1(\Omega))^3$ onto $\mathbf{V}_j^{div,0}$. Again, for all $\boldsymbol{u} \in (H^s(\Omega))^3$ with $1 \le s \le r-1$, the following Jackson type estimation holds, for some $C > 0$:

$$\|\boldsymbol{u} - \mathbb{P}_j^{div,0}(\boldsymbol{u})\|_{H_0^1} \le C\, 2^{-j(s-1)} \|\boldsymbol{u}\|_{H^s}. \tag{24}$$

2.3 *Extension to Higher Dimension d > 3*

The divergence-free wavelet basis construction is not limited to dimension 3 only. Several applications, such as image warping [29], optimal transportation [18], or magnetohydrodynamic turbulence [12], involve divergence-free vector fields living in spaces of dimension d greater than 3. This divergence-free constraint for the solution leads in general to solve a Poisson equation difficult to handle. Therefore, it would be of great interest to have at hand divergence-free bases that enable to encode such solution.

The construction of previous section extends to larger dimensions $d > 3$ readily. As in [10, 25], we obtain in this case $(d-1)$ types of linear independent divergence-free wavelet functions. For $1 \leq i \leq d-1$, the general formula of these wavelets is given by

$$\Psi_{j,k}^{div,i} := \begin{bmatrix} 0 \\ \vdots \\ 0 \\ 2^{j_{i+1}} \psi_{j_1,k_1}^0 \otimes \cdots \otimes \psi_{j_{i-1},k_{i-1}}^0 \otimes \psi_{j_i,k_i}^d \otimes \psi_{j_{i+1},k_{i+1}}^0 \otimes \cdots \otimes \psi_{j_d,k_d}^0 \\ -2^{j_i} \psi_{j_1,k_1}^0 \otimes \cdots \otimes \psi_{j_i,k_i}^0 \otimes \psi_{j_{i+1},k_{i+1}}^d \otimes \psi_{j_{i+1},k_{i+2}}^0 \otimes \cdots \otimes \psi_{j_d,k_d}^0 \\ 0 \\ \vdots \\ 0 \end{bmatrix}, \tag{25}$$

where only the row i and row $i+1$ are not zeros.

Recalling that $(\psi_{j_i,k_i}^d)' = 2^{j_i} \psi_{j_i,k_i}^0$, we easily verify that $\nabla \cdot \Psi_{j,k}^{div,i} = 0$, and that $\Psi_{j,k}^{div,i}$ satisfy the boundary condition $\Psi_{j,k}^{div,i} \cdot \boldsymbol{n} = 0$. The space $\mathbf{W}_j^{div}$ spanned by these wavelets is included in the following standard BMRA of $(L^2(\Omega))^d$:

$$\mathbf{V}_j = \mathbf{V}_j^1 \times \cdots \times \mathbf{V}_j^d \quad \text{with} \quad \mathbf{V}_j^i = V_j^{\delta_{1,i}} \otimes \cdots \otimes V_j^{\delta_{d,i}}, \quad 1 \leq i \leq d, \tag{26}$$

where $\delta_{i,j}$ denotes the Kronecker symbol. To satisfy the free-slip boundary condition, we must replace V_j^1 by V_j^d in (26). We also emphasized that the corresponding spaces $\mathbf{V}_j^{div} = \mathbf{V}_j \cap \mathcal{H}_{div}(\Omega) = \mathbf{P}_j(\mathcal{H}_{div}(\Omega))$ provide a multiresolution analysis of $\mathcal{H}_{div}(\Omega)$. Following a similar approach as for $d = 3$, and taking $(d-1)$ scaling functions, we obtain a divergence-free basis of $\mathbf{V}_j^{div}$.

3 Divergence-Free Wavelet Schemes for the Navier–Stokes Equations

The motions of incompressible homogeneous viscous flows, confined in an open domain $\Omega \subset \mathbb{R}^d$ with smooth boundary $\Gamma = \partial\Omega$, are governed by the time-depending Navier–Stokes equations:

$$\begin{cases} \partial_t \boldsymbol{v} - \nu \Delta \boldsymbol{v} + (\boldsymbol{v} \cdot \nabla)\boldsymbol{v} + \nabla p = \boldsymbol{f}, \\ \nabla \cdot \boldsymbol{v} = 0, \end{cases} \tag{27}$$

where $\boldsymbol{v} \in \mathbb{R}^d$ denotes the velocity vector field, $p \in \mathbb{R}$ is the pressure, $\nu > 0$ is the kinematic viscosity, and $\boldsymbol{f}$ is the external force. Without loss of generality, we will assume that $\boldsymbol{f} = 0$. System (27) is supplemented by boundary conditions, and throughout this section, we will assume the no-slip boundary condition $\boldsymbol{v} = 0$ on Γ. Non-homogeneous Dirichlet boundary condition $\boldsymbol{v} = g$ on Γ will be handled in the numerical experiments.

The objective in this section is to study the numerical approximation of (27) using divergence-free wavelet-based methods. These schemes use the divergence-free wavelet Leray projector to decouple the computation of the velocity and the pressure, as in the case of the Fourier spectral method [8, 9, 22]; thus, they avoid the resolution of a Stokes problem. In this section, we propose to investigate the generalization of the method of [22] to the three-dimensional Navier–Stokes equations. We begin with the description and the study of the temporal and spatial discretizations.

3.1 Temporal Discretization

Let $\delta t > 0$ be a time step. For $0 \leq n \leq N$, we set $t_n = n\delta t$ such that $0 = t_0 < t_1 < \cdots < t_N = T$ is a uniform partition of the computational time interval. We denote by $\boldsymbol{v}^n$ and p^n the approximations of $\boldsymbol{v}(t_n, x)$ and $p(t_n, x)$, where $(\boldsymbol{v}, p)$ is a smooth solution of (27). To compute $(\boldsymbol{v}^n, p^n)$, the approach of [22] consists in replacing the classical correction step of the projection method [4, 24, 33] by a divergence-free wavelet Leray–Hodje projector. Specifically, since $(H_0^1(\Omega))^d = \mathcal{H}_{div,0}(\Omega) \oplus \mathcal{H}_{div,0}(\Omega)^{\perp}$, there exists Φ^{n+1}, a scalar potential in $L^2(\Omega)$ such that

$$\tilde{\boldsymbol{v}}^{n+1} = \boldsymbol{v}^{n+1} + \nabla\Phi^{n+1}, \quad \text{with } \tilde{\boldsymbol{v}}^{n+1} \in (H_0^1(\Omega))^d. \tag{28}$$

Then, substituting this change of variable in the Navier–Stokes equations (27) and using a projection method time step, with an implicit Euler scheme in time for the diffusion term, we get

- Prediction step:

$$\begin{cases} \frac{\tilde{\boldsymbol{v}}^{n+1} - \boldsymbol{v}^n}{\delta t} + (\boldsymbol{v}^n \cdot \nabla)\boldsymbol{v}^n = \nu\Delta\tilde{\boldsymbol{v}}^{n+1}, \\ \tilde{\boldsymbol{v}}^{n+1} = 0, \quad \text{on } \partial\Omega. \end{cases} \tag{29}$$

- Correction step:

$$\begin{cases} \boldsymbol{v}^{n+1} = \mathbb{P}^{div,0}(\tilde{\boldsymbol{v}}^{n+1}), \\ \\ p^{n+1} = \frac{1}{\delta t}\Phi^{n+1} - \nu\Delta\Phi^{n+1}, \end{cases} \tag{30}$$

where the correction step of the standard projection method has been modified by introducing $\mathbb{P}^{div,0}$, the L^2-orthogonal projector from $(H_0^1(\Omega))^2$ onto $\mathcal{H}_{div,0}(\Omega)$. Remark that the Navier–Stokes formulation (29) and (30) is no more than a change of variables, whereas the classical projection method is an operator splitting, which implies a loss of precision in time.

The problem defined by (29) and (30) is well posed, in the sense that the numerical solution $\tilde{\boldsymbol{v}}^{n+1}$ exists for given smooth initial data. Indeed, let us define the bilinear form $a(.,.)$:

$$a(\boldsymbol{u}, \boldsymbol{v}) = \int_\Omega \boldsymbol{u} \cdot \boldsymbol{v} + \nu\delta t \int_\Omega \nabla\boldsymbol{u} : \nabla\boldsymbol{v}, \quad \forall\, \boldsymbol{u}, \boldsymbol{v} \in H_0^1(\Omega)^d, \tag{31}$$

and the linear form L:

$$L(\boldsymbol{v}) = \int_\Omega \boldsymbol{v}^n \cdot \boldsymbol{v} - \delta t \int_\Omega (\boldsymbol{v}^n \cdot \nabla)\boldsymbol{v}^n \cdot \boldsymbol{v}, \forall\, \boldsymbol{v} \in H_0^1(\Omega)^d. \tag{32}$$

Formal computations lead to

$$a(\boldsymbol{v}, \boldsymbol{v}) = \int_\Omega \boldsymbol{v} \cdot \boldsymbol{v} + \nu\delta t \int_\Omega |\nabla\boldsymbol{v}|^2 \geq \min\{1, \nu\delta t\}\, \|\boldsymbol{v}\|_{H^1}^2, \quad \forall \boldsymbol{v} \in H_0^1(\Omega)^d,$$

and

$$|L(\boldsymbol{v})| \leq \left(\|\boldsymbol{v}^n\|_{H^1} + \delta t\, \|\boldsymbol{v}^n\|_{L^\infty} \|\boldsymbol{v}^n\|_{H^1}\right) \|\boldsymbol{v}\|_{H^1}, \quad \forall\, \boldsymbol{v} \in H_0^1(\Omega)^d.$$

By the Lax–Milgram theorem [16, 34], there exists a unique $\tilde{\boldsymbol{v}}^{n+1} \in H_0^1(\Omega)^d$, solution of (29) and $\boldsymbol{v}^{n+1} = \mathbb{P}^{div,0}(\tilde{\boldsymbol{v}}^{n+1})$. The spatial approximation of these solutions is detailed in next subsection.

3.2 Spatial Discretization

For simplicity, this part takes place in dimension $d = 3$ but can be generalized in general dimension. At a fixed spatial resolution $j \geq j_{min}$, the numerical solutions

$\tilde{v}^n$ and v^n of (29-30) are searched as the following linear combination of wavelets:

$$\tilde{v}^n_j = \sum_{\epsilon=1}^{3} \left(\sum_{j_{min}-1 \le |\boldsymbol{j}| < j} \sum_{\boldsymbol{k}} \tilde{d}^{n,\epsilon}_{\boldsymbol{j},\boldsymbol{k}} \Psi^{\epsilon}_{\boldsymbol{j},\boldsymbol{k}} \right) \text{ and } v^n_j = \sum_{\epsilon=1}^{2} \left(\sum_{j_{min}-1 \le |\boldsymbol{j}| < j} \sum_{\boldsymbol{k}} d^{div,n,\epsilon}_{\boldsymbol{j},\boldsymbol{k}} \Psi^{div,0,\epsilon}_{\boldsymbol{j},\boldsymbol{k}} \right),$$

where for $\epsilon = 1, 2, 3$, $\Psi^{\epsilon}_{\boldsymbol{j},\boldsymbol{k}}$ denote the usual tensor product wavelets of the space $\mathbf{V}^{dd}_j$ introduced in (20), whose (vector) scaling functions $\Phi^{\epsilon}_{j,\boldsymbol{k}}$ are recalled below:

$$\Phi^1_{j,\boldsymbol{k}} = \begin{bmatrix} \varphi^d_{j,k_1} \otimes \varphi^{00}_{j,k_2} \otimes \varphi^{00}_{j,k_3} \\ 0 \\ 0 \end{bmatrix}, \Phi^2_{j,\boldsymbol{k}} = \begin{bmatrix} 0 \\ \varphi^{00}_{j,k_1} \otimes \varphi^d_{j,k_2} \otimes \varphi^{00}_{j,k_3} \\ 0 \end{bmatrix}, \Phi^3_{j,\boldsymbol{k}} = \begin{bmatrix} 0 \\ 0 \\ \varphi^{00}_{j,k_1} \otimes \varphi^{00}_{j,k_2} \otimes \varphi^d_{j,k_3} \end{bmatrix}.$$

For the wavelet basis $\Psi^{\epsilon}_{\boldsymbol{j},\boldsymbol{k}}$, we adopted the convention that at index $j_{min} - 1$, the wavelets have to be replaced by scaling functions:

$$\Psi^{\epsilon}_{j_{min}-1,\boldsymbol{k}} = \Phi^{\epsilon}_{j_{min},\boldsymbol{k}} \quad \text{and} \quad \Psi^{div,0,\epsilon}_{j_{min}-1,\boldsymbol{k}} = \Phi^{div,0,\epsilon}_{j_{min},\boldsymbol{k}}.$$

We use a Galerkin method to compute the set of coefficients $(\tilde{d}^{n+1,\epsilon}_{\boldsymbol{j},\boldsymbol{k}})$ and $(d^{div,n+1,\epsilon}_{\boldsymbol{j},\boldsymbol{k}})$ on these wavelet bases. The numerical resolution of (29-30) thus yields to the resolution of two linear systems:

$$\mathbb{M}_j(\tilde{d}^{n+1,\epsilon}_{\boldsymbol{j},\boldsymbol{k}}) = \mathbb{A}_j \left((d^{n,\epsilon}_{\boldsymbol{j},\boldsymbol{k}}) + \delta t (h^{n,\epsilon}_{\boldsymbol{j},\boldsymbol{k}}) \right), \tag{33}$$

and

$$\mathbb{M}^{div}_j(d^{div,n+1,\epsilon}_{\boldsymbol{j},\boldsymbol{k}}) = \left((\langle \tilde{v}^{n+1}_j, \Psi^{div,0,\epsilon}_{\boldsymbol{j},\boldsymbol{k}} \rangle) \right), \tag{34}$$

where $d^{n,\epsilon}_{\boldsymbol{j},\boldsymbol{k}}$ and $h^{n,\epsilon}_{\boldsymbol{j},\boldsymbol{k}}$ are, respectively, the coefficients of the projection of v^n and $(v^n \cdot \nabla)v^n$ onto the basis $\{\Psi^{\epsilon}_{\boldsymbol{j},\boldsymbol{k}}\}$. The nonlinear term $(v^n \cdot \nabla)v^n$ is computed at grid collocation points before its projection onto the wavelet space, where the gradient operator ∇ is approximated using a finite difference method of the same order as the scaling function φ^1 polynomial approximation order [22]. Likewise, $\mathbb{A}_j$ and $\mathbb{M}_j$ correspond, respectively, to the matrices of the projection of the identity operator and $1 - \delta t \Delta$ onto this wavelet space, and $\mathbb{M}^{div}_j$ denotes the divergence-free wavelet basis Gramian matrix. In practice, one can take advantage of the tensor product construction of the above wavelet bases, thus reducing the storage and the numerical complexity of vector-matrix multiplication in (33) and (34). For the numerical computation and properties of these matrices and projections, the reader is referred to [1, 20, 22].

The numerical schemes (29–30) and (33–34) are stable and consistent with the Navier–Stokes equations (27). This can be deduced from the numerical error estimations.

3.3 Numerical Error Estimations

Let v_j^n be the numerical solution of (33) and (34). If the initial condition v^0 is regular enough, the following lemma is verified:

Lemma 1 *For v_j^n an approximated solution of* (30) *given by* (33) *and* (34)*, with an appropriated time step δt, we have*

$$\|v_j^n\|_{L^2}^2 \leq C(v^0, \nu) \tag{35}$$

and

$$\nu\delta t \sum_{k=1}^{n} \|\nabla v_j^k\|_{L^2}^2 \leq C(v^0, \nu), \tag{36}$$

where $C(v^0, \nu)$ is a positive constant depending only on the initial data.

Proof. To prove the lemma, it suffices to show that

$$\|v_j^n\|_{L^2}^2 + \nu\delta t C \sum_{k=1}^{n} \|\nabla v_j^k\|_{L^2}^2 \leq C(v^0, \nu), \tag{37}$$

and this is done by induction with similar arguments as in the proof of Lemma 5.9 of [34]. Let us introduce a constant $C(v^0, \nu)$ such that

$$C(v^0, \nu) = \|v^0\|_{L^2}^2 + \delta t^2 2^{jd} \|v^0\|_{L^2}^2 \|\nabla v^0\|_{L^2}^2. \tag{38}$$

Now, we assume that (37) holds for some $n \geq 0$. Then, using (31) and (32), the variational formulation for the solution $\tilde{v}_j^{n+1}$ reads

$$a(\tilde{v}_j^{n+1}, v) = L(v), \quad \forall\, v \in \mathbf{V}_j^d, \quad \text{for } j > j_{min}. \tag{39}$$

Since $\nabla \cdot v_j^{n+1} = 0$, taking $2v_j^{n+1}$ as a test function in (39) and replacing $\tilde{v}_j^{n+1}$ by its exact expression in (28) leads to

$$\|v_j^{n+1}\|_{L^2}^2 - \|v_j^n\|_{L^2}^2 + \|v_j^{n+1} - v_j^n\|_{L^2}^2 + 2\nu\delta t \|\nabla v_j^{n+1}\|_{L^2}^2 = -2\delta t \langle (v_j^n \cdot \nabla) v_j^n, v_j^{n+1} \rangle. \tag{40}$$

Otherwise, we have

$$\begin{aligned}\delta t \langle (v_j^n \cdot \nabla) v_j^n, v_j^{n+1} \rangle = \delta t \langle (v_j^n \cdot \nabla) v_j^n, v_j^{n+1} - v_j^n \rangle &\leq \delta t \|v_j^n\|_{L^\infty} \|\nabla v_j^n\|_{L^2} \|v_j^{n+1} - v_j^n\|_{L^2} \\ &\leq \frac{\delta t^2}{2} \|v_j^n\|_{L^\infty}^2 \|\nabla v_j^n\|_{L^2}^2 + \frac{1}{2} \|v_j^{n+1} - v_j^n\|_{L^2}^2.\end{aligned}$$

Thus, collecting these estimations together, we deduce that

$$\|\boldsymbol{v}_j^{n+1}\|_{L^2}^2 - \|\boldsymbol{v}_j^n\|_{L^2}^2 + 2\nu\delta t\|\nabla \boldsymbol{v}_j^{n+1}\|_{L^2}^2 \le \delta t^2 \|\boldsymbol{v}_j^n\|_{L^\infty}^2 \|\nabla \boldsymbol{v}_j^n\|_{L^2}^2. \tag{41}$$

We recall that $\boldsymbol{v}_j^n \in \mathbf{V}_j^d$ for $j > j_{min}$. As we are in finite space dimension, we have

$$\|\boldsymbol{v}_j^n\|_{L^\infty}^2 \le 2^{jd} \|\boldsymbol{v}_j^n\|_{L^2}^2, \tag{42}$$

and Eq. (41) becomes

$$\|\boldsymbol{v}_j^{n+1}\|_{L^2}^2 - \|\boldsymbol{v}_j^n\|_{L^2}^2 + 2\nu\delta t\|\nabla \boldsymbol{v}_j^{n+1}\|_{L^2}^2 \le \delta t^2 2^{jd} \|\boldsymbol{v}_j^n\|_{L^2}^2 \|\nabla \boldsymbol{v}_j^n\|_{L^2}^2. \tag{43}$$

For $n = 1$, (41) reads

$$\|\boldsymbol{v}_j^1\|_{L^2}^2 + 2\nu\delta t\|\nabla \boldsymbol{v}_j^1\|_{L^2}^2 \le \|\boldsymbol{v}_j^0\|_{L^2}^2 + \delta t^2 2^{jd} \|\boldsymbol{v}_j^0\|_{L^2}^2 \|\nabla \boldsymbol{v}_j^0\|_{L^2}^2. \tag{44}$$

This is the statement of the lemma with

$$C(\boldsymbol{v}^0, \nu) \ge \|\boldsymbol{v}_j^0\|_{L^2}^2 + \delta t^2 2^{jd} \|\boldsymbol{v}_j^0\|_{L^2}^2 \|\nabla \boldsymbol{v}_j^0\|_{L^2}^2. \tag{45}$$

Summation over n in (41) leads to

$$\|\boldsymbol{v}_j^{n+1}\|_{L^2}^2 - \|\boldsymbol{v}_j^0\|_{L^2}^2 + 2\nu\delta t \sum_{k=0}^{n} \|\nabla \boldsymbol{v}_j^{k+1}\|_{L^2}^2 \le \delta t^2 2^{jd} \sum_{k=0}^{n} \|\boldsymbol{v}_j^k\|_{L^2}^2 \|\nabla \boldsymbol{v}_j^k\|_{L^2}^2, \tag{46}$$

and rewritten this, we get

$$\|\boldsymbol{v}_j^{n+1}\|_{L^2}^2 + 2\nu\delta t\|\nabla \boldsymbol{v}_j^{n+1}\|_{L^2}^2 + 2\nu\delta t \sum_{k=1}^{n} \|\nabla \boldsymbol{v}_j^k\|_{L^2}^2 \le C(\boldsymbol{v}^0, \nu) + \delta t^2 2^{jd} C(\boldsymbol{v}^0, \nu) \sum_{k=1}^{n} \|\nabla \boldsymbol{v}_j^k\|_{L^2}^2,$$

which implies

$$\|\boldsymbol{v}_j^{n+1}\|_{L^2}^2 + 2\nu\delta t\|\nabla \boldsymbol{v}_j^{n+1}\|_{L^2}^2 + \delta t[2\nu - \delta t 2^{jd} C(\boldsymbol{v}^0, \nu)] \sum_{k=1}^{n} \|\nabla \boldsymbol{v}_j^k\|_{L^2}^2 \le C(\boldsymbol{v}^0, \nu). \tag{47}$$

From (47), to end the proof of the lemma, we take δt small enough such that

$$\delta t 2^{jd} C(\boldsymbol{v}^0, \nu) < 2\nu. \tag{48}$$

□

Now, we will analyze the behavior of the local error $\boldsymbol{e}^n = \boldsymbol{v}(t_n) - \boldsymbol{v}^n$, where $\boldsymbol{v}$ is a smooth solution of the exact Navier–Stokes equations (27) and $\boldsymbol{v}^n$ the numerical solution of and (30). With the help of the Lemma 1, we get

Proposition 2 *Let $\boldsymbol{v} \in H^2(0,T;L^2(\Omega)^d) \cap C^0(0,T;W^{1,+\infty}(\Omega)^d \cap H_0^1(\Omega)^d)$ be a solution of (27) and $\boldsymbol{v}^n$ a solution of (29 -30). Denoting by $\boldsymbol{e}^n = \boldsymbol{v}(t_n) - \boldsymbol{v}^n$ the local error, we have*

$$\max_n \|\boldsymbol{e}^n\|^2_{L^2(\Omega)^d} \xrightarrow[\delta t \to 0]{} 0. \tag{49}$$

Proof. According to (29) and (30), we see that

$$\boldsymbol{v}(t_{n+1}) - \boldsymbol{v}(t_n) - \delta t \nu \Delta \boldsymbol{v}(t_{n+1}) + \delta t(\boldsymbol{v}(t_n)\cdot\nabla)\boldsymbol{v}(t_n) = \delta t \boldsymbol{\epsilon}^{n+1} - \delta t \nabla p(t_n), \tag{50}$$

where $\boldsymbol{\epsilon}^n$ denotes the consistency error. Since $\nabla \cdot \boldsymbol{e}^{n+1} = 0$, taking $2\boldsymbol{e}^{n+1}$ as test function in the variational formulations and using similar arguments as in the proof of Lemma 1 allow to get

$$\begin{aligned}&\|\boldsymbol{e}^{n+1}\|^2_{L^2} - \|\boldsymbol{e}^n\|^2_{L^2} + \|\boldsymbol{e}^{n+1} - \boldsymbol{e}^n\|^2_{L^2} + 2\delta t \nu \|\nabla \boldsymbol{e}^{n+1}\|^2_{L^2}\\ &\quad\le 2\delta t \|\boldsymbol{v}^n\|_{L^\infty}\|\nabla \boldsymbol{e}^n\|_{L^2}\|\boldsymbol{e}^{n+1} - \boldsymbol{e}^n\|_{L^2} + 2\delta t\|\boldsymbol{e}^n\|_{L^2}\|\nabla \boldsymbol{v}(t_n)\|_{L^\infty}\|\boldsymbol{e}^{n+1}\|_{L^2}\\ &\quad+ 2\delta t\langle \boldsymbol{\epsilon}^{n+1}, \boldsymbol{e}^{n+1}\rangle. \end{aligned}\tag{51}$$

Poincaré and Young's inequalities lead to

$$\begin{aligned}\|\boldsymbol{e}^{n+1}\|^2_{L^2} - \|\boldsymbol{e}^n\|^2_{L^2} + \delta t \nu \|\nabla \boldsymbol{e}^{n+1}\|^2_{L^2} &\le \delta t^2 \|\boldsymbol{v}^n\|^2_{L^\infty}\|\nabla \boldsymbol{e}^n\|^2_{L^2}\\ &+ \frac{C(\Omega)^2\delta t}{\nu}\|\boldsymbol{e}^n\|^2_{L^2}\|\nabla \boldsymbol{v}(t_n)\|^2_{L^\infty} + 2\delta t\langle \boldsymbol{\epsilon}^{n+1}, \boldsymbol{e}^{n+1}\rangle.\end{aligned}\tag{52}$$

Since $\nabla \cdot \boldsymbol{e}^n = 0$, using (50), we have

$$\delta t\langle \boldsymbol{\epsilon}^{n+1}, \boldsymbol{e}^{n+1}\rangle = \langle \boldsymbol{v}(t_{n+1}) - \boldsymbol{v}(t_n) - \delta t \nu \Delta \boldsymbol{v}(t_{n+1}) + \delta t(\boldsymbol{v}(t_n)\cdot\nabla)\boldsymbol{v}(t_n), \boldsymbol{e}^{n+1}\rangle, \tag{53}$$

and replacing $-\Delta \boldsymbol{v}(t_{n+1})$ by

$$-\nu\Delta \boldsymbol{v}(t_{n+1}) = -\partial_t \boldsymbol{v}(t_{n+1}) - (\boldsymbol{v}(t_{n+1})\cdot\nabla)\boldsymbol{v}(t_{n+1}) - \nabla p(t_{n+1}), \tag{54}$$

we obtain

$$\begin{aligned}&\delta t\langle \boldsymbol{\epsilon}^{n+1}, \boldsymbol{e}^{n+1}\rangle\\ &= \langle \boldsymbol{v}(t_{n+1}) - \boldsymbol{v}(t_n) - \delta t \partial_t \boldsymbol{v}(t_{n+1}) + \delta t[(\boldsymbol{v}(t_n)\cdot\nabla)\boldsymbol{v}(t_n) - (\boldsymbol{v}(t_{n+1})\cdot\nabla)\boldsymbol{v}(t_{n+1})], \boldsymbol{e}^{n+1}\rangle\\ &= \delta t\langle \frac{1}{\delta t}\int_{t_n}^{t_{n+1}} (t_n - t)\partial_{tt}\boldsymbol{v}(t)dt - \int_{t_n}^{t_{n+1}} \partial_t[(\boldsymbol{v}(t)\cdot\nabla)\boldsymbol{v}(t)]dt, \boldsymbol{e}^{n+1}\rangle\\ &\le \left(\frac{\delta t^{3/2}}{\sqrt{3}}\|\partial_{tt}\boldsymbol{v}\|_{L^2(t_n,t_{n+1};L^2)} + \delta t^{3/2}\|\partial_t[(\boldsymbol{v}\cdot\nabla)\boldsymbol{v}]\|_{L^2(t_n,t_{n+1};L^2)}\right)\|\boldsymbol{e}^{n+1}\|_{L^2}\end{aligned}\tag{55}$$

Again, using Young's inequality, we have

$$2\delta t\langle \boldsymbol{\epsilon}^{n+1}, \boldsymbol{e}^{n+1}\rangle \leq \frac{4C(\Omega)^2\delta t^2}{3\nu}\|\partial_{tt}\boldsymbol{v}\|^2_{L^2(t_n,t_{n+1};L^2)} + \frac{4C(\Omega)^2\delta t^2}{\nu}\|\partial_t[(\boldsymbol{v}\cdot\nabla)\boldsymbol{v}]\|^2_{L^2(t_n,t_{n+1};L^2)} + \frac{\nu\delta t}{2}\|\nabla \boldsymbol{e}^{n+1}\|^2_{L^2},$$

and from the maximum principle and regularity of the solution of elliptic problem, we infer that

$$\|\boldsymbol{v}^n\|_{L^\infty} \lesssim \|\boldsymbol{v}^n_j\|_{L^\infty}.$$

Thus, (52) becomes

$$\|\boldsymbol{e}^{n+1}\|^2_{L^2} - \|\boldsymbol{e}^n\|^2_{L^2} + \frac{\delta t\nu}{2}\|\nabla\boldsymbol{e}^{n+1}\|^2_{L^2} \leq \delta t^2 2^{jd}\|\boldsymbol{v}^n_j\|^2_{L^2}\|\nabla\boldsymbol{e}^n\|^2_{L^2} + \frac{C(\Omega)^2\delta t}{\nu}\|\boldsymbol{e}^n\|^2_{L^2}\|\nabla\boldsymbol{v}(t_n)\|^2_{L^\infty}$$
$$+ \frac{4C(\Omega)^2\delta t^2}{3\nu}\|\partial_{tt}\boldsymbol{v}\|^2_{L^2(t_n,t_{n+1};L^2)}$$
$$+ \frac{4C(\Omega)^2\delta t^2}{\nu}\|\partial_t[(\boldsymbol{v}\cdot\nabla)\boldsymbol{v}]\|^2_{L^2(t_n,t_{n+1};L^2)}.$$

Summation over n in (52) shows that

$$\|\boldsymbol{e}^{n+1}\|^2_{L^2} - \|\boldsymbol{e}^0\|^2_{L^2} + \frac{\delta t\nu}{2}\sum_{k=1}^{n+1}\|\nabla\boldsymbol{e}^k\|^2_{L^2} \leq \delta t^2 2^{jd}\sum_{k=0}^{n}\|\boldsymbol{v}^k_j\|^2_{L^2}\|\nabla\boldsymbol{e}^k\|^2_{L^2}$$
$$+C_1\delta t\sum_{k=0}^{n}\|\boldsymbol{e}^k\|^2_{L^2} + C_2\delta t^2,$$

with

$$C_1 = \frac{C(\Omega)^2}{\nu}\|\nabla\boldsymbol{v}\|^2_{L^\infty(0,T,L^\infty)},\ C_2 = \frac{4C(\Omega)^2}{3\nu}\|\partial_{tt}\boldsymbol{v}\|^2_{L^2(0,T;L^2)} + \frac{4C(\Omega)^2}{\nu}\|\partial_t[(\boldsymbol{v}\cdot\nabla)\boldsymbol{v}]\|^2_{L^2(0,T;L^2)}.$$

Since $\|\boldsymbol{v}^k_j\|^2_{L^2} \leq C(\boldsymbol{v}^0,\nu)$ according to Lemma 1, we have

$$\|\boldsymbol{e}^{n+1}\|^2_{L^2} + \frac{\delta t\nu}{2}\|\nabla\boldsymbol{e}^{n+1}\|^2_{L^2} + \left(\frac{\delta t\nu}{2} - \delta t^2 2^{jd}C(\boldsymbol{v}^0,\nu)\right)\sum_{k=1}^{n}\|\nabla\boldsymbol{e}^k\|^2_{L^2}$$
$$\leq \lambda_0 + C_1\delta t\sum_{k=0}^{n}\|\boldsymbol{e}^k\|^2_{L^2} + C_2\delta t^2,$$

where

$$\lambda_0 = \|\boldsymbol{e}^0\|^2_{L^2} + \delta t^2 2^{jd}C(\boldsymbol{v}^0,\nu)\|\nabla\boldsymbol{e}^0\|^2_{L^2}.$$

Then, choosing δt small enough such that

$$\frac{\nu}{2} - \delta t 2^{jd} C(\boldsymbol{v}^0, \nu) > 0, \tag{56}$$

by the discrete Gronwall lemma, see [34, 37], we deduce that

$$\|\boldsymbol{e}^n\|_{L^2}^2 \leq \left(\lambda_0 + C_2 \delta t^2\right) \exp(C_1 T) \tag{57}$$

□

For the numerical error $\boldsymbol{e}_j^n = \boldsymbol{v}(t_n) - \boldsymbol{v}_j^n$, we can write

$$\|\boldsymbol{e}_j^n\|_{L^2}^2 \leq 2\left(\|\boldsymbol{v}(t_n) - \boldsymbol{v}^n\|_{L^2}^2 + \|\boldsymbol{v}^n - \boldsymbol{v}_j^n\|_{L^2}^2\right),$$

and

$$\|\boldsymbol{v}^n - \boldsymbol{v}_j^n\|_{L^2}^2 \leq C\|\boldsymbol{v}^n - \mathbb{P}_j^{div,0}(\boldsymbol{v}^n)\|_{L^2}^2.$$

Thus, due to the Jackson estimation (24) and Proposition 2, we obtain the following convergence result:

$$\|\boldsymbol{e}_j^n\|_{L^2(\Omega)^d}^2 \xrightarrow[\delta t \to 0, \delta x \to 0]{} 0. \tag{58}$$

4 Numerical Results

We present in this section numerical results on the simulation of the Navier–Stokes equations, obtained using the time and spatial discretization (29-30) and (33-34). These results will be compared to those of the literature for method validation.

4.1 Divergence Free Wavelet Illustration

We start by presenting the wavelet bases generators introduced in Sect. 2.1, used in all the numerical experiments that follow. We chose as scaling function generators $(\varphi^1, \tilde{\varphi}^1)$, the biorthogonal splines of order 4: $r = \tilde{r} = 4$. This corresponds to four vanishing moments for each wavelet generator ψ^1 and $\tilde{\psi}^1$. We plot in Fig. 1 the graph of φ^1 and its derivative that satisfies $(\varphi^1(x))' = \varphi^0(x) - \varphi^0(x-1)$. Thus, the scaling function φ^0 can reproduce polynomials up to order 3 (or equivalently

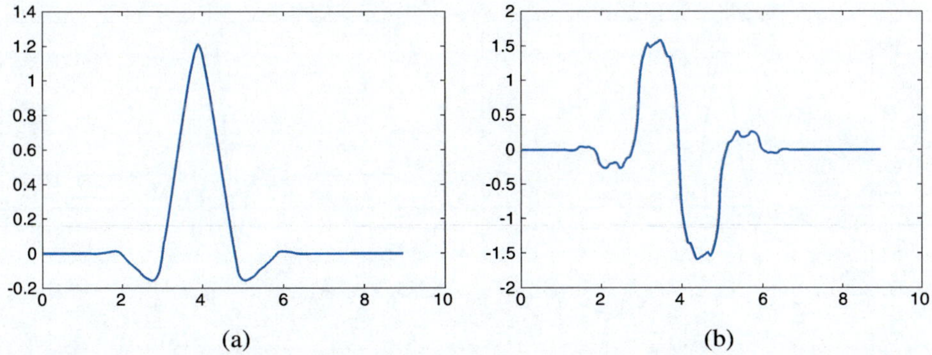

Fig. 1 Primal biorthogonal spline scaling function generator φ^1 (left) and its derivative $(\varphi^1(x))' = \varphi^0(x) - \varphi^0(x-1)$ (right) with order parameters $r = \tilde{r} = 4$. (**a**) φ^1. (**b**) $(\varphi^1)'$

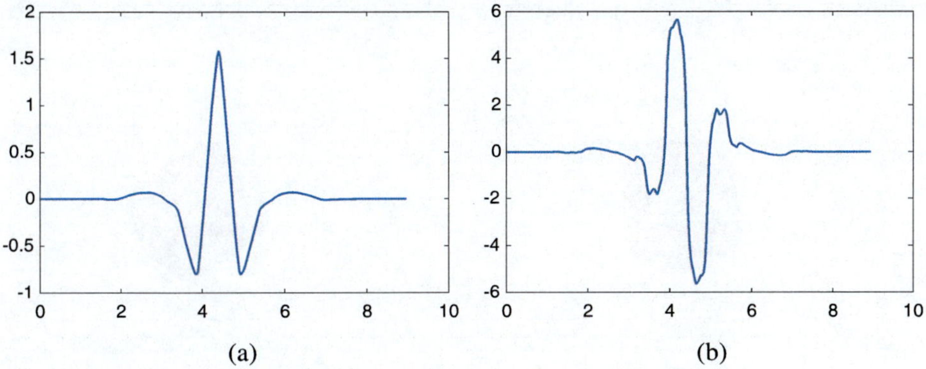

Fig. 2 Primal biorthogonal spline wavelet generator ψ^1 (left) and its derivative $(\psi^1(x))' = 4\psi^0(x)$ (right) with order parameters $r = \tilde{r} = 4$. (**a**) ψ^1. (**b**) $4\psi^0$

of degree up to 2). Figure 2 shows the graph of the wavelet generators ψ^1 and $(\psi^1)' = 4\psi^0$. The biorthogonal functions $(\tilde{\varphi}^1, \tilde{\psi}^1)$ are plotted in Fig. 3.

As mentioned in Sect. 2.2, the above scaling functions allow to construct, in dimension 3, three divergence-free scaling function generators, defined by (10–12).

We plot in Fig. 4 the isosurface of the magnitude of some internal scaling functions $\Phi^{div,1}_{j,\mathbf{k}}$, $\Phi^{div,2}_{j,\mathbf{k}}$, and $\Phi^{div,3}_{j,\mathbf{k}}$, where for a vector function $\boldsymbol{v} = (v_1, v_2, v_3)$, the magnitude is defined as $|\boldsymbol{v}| = \sqrt{v_1^2 + v_2^2 + v_3^2}$. Likewise, the corresponding divergence-free wavelet generators magnitude isosurfaces are shown in Fig. 5.

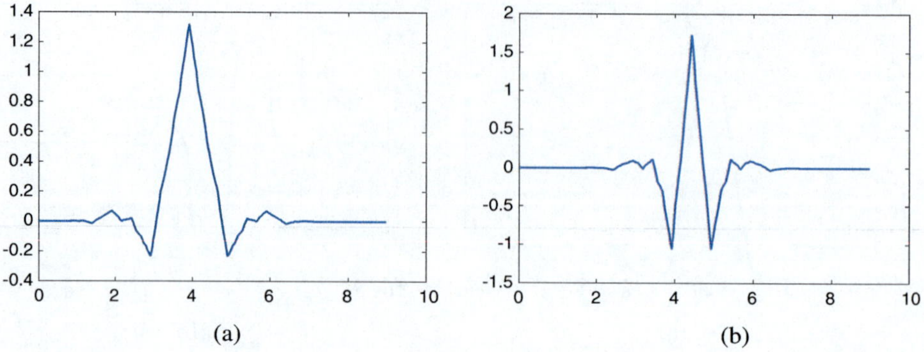

Fig. 3 Dual biorthogonal spline scaling function generator $\tilde{\varphi}^1$ (left) and the dual wavelet generator $\tilde{\psi}^1$ (right) with order parameters $r = \tilde{r} = 4$. **(a)** $\tilde{\varphi}^1$. **(b)** $\tilde{\psi}^1$

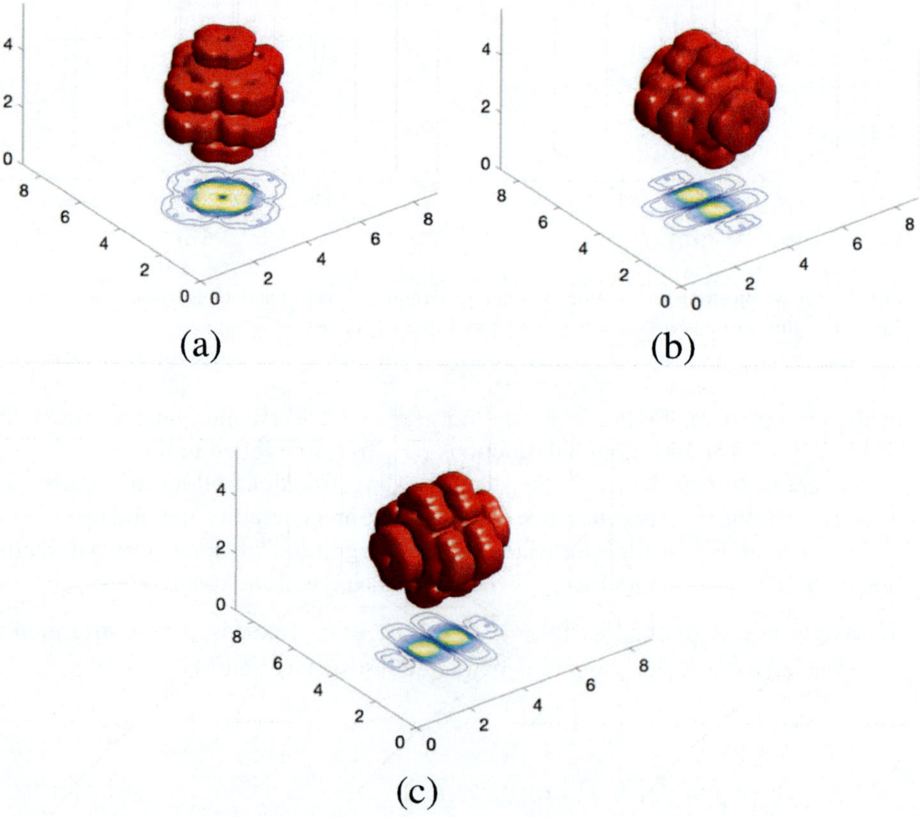

Fig. 4 Divergence-free scaling functions magnitude isosurface for biorthogonal spline generators $(\varphi^1, \tilde{\varphi}^1)$ of order 4. **(a)** $|\Phi^{div,1}_{j,\mathbf{k}}| = 0.2$. **(b)** $|\Phi^{div,2}_{j,\mathbf{k}}| = 0.2$. **(c)** $|\Phi^{div,3}_{j,\mathbf{k}}| = 0.2$

4.2 Analyses of Time and Space Convergence Rates

In this part, we study the time and space convergence rates provided by the divergence-free wavelet-based projection method (29–30)–(33–34) introduced in Sect. 3.

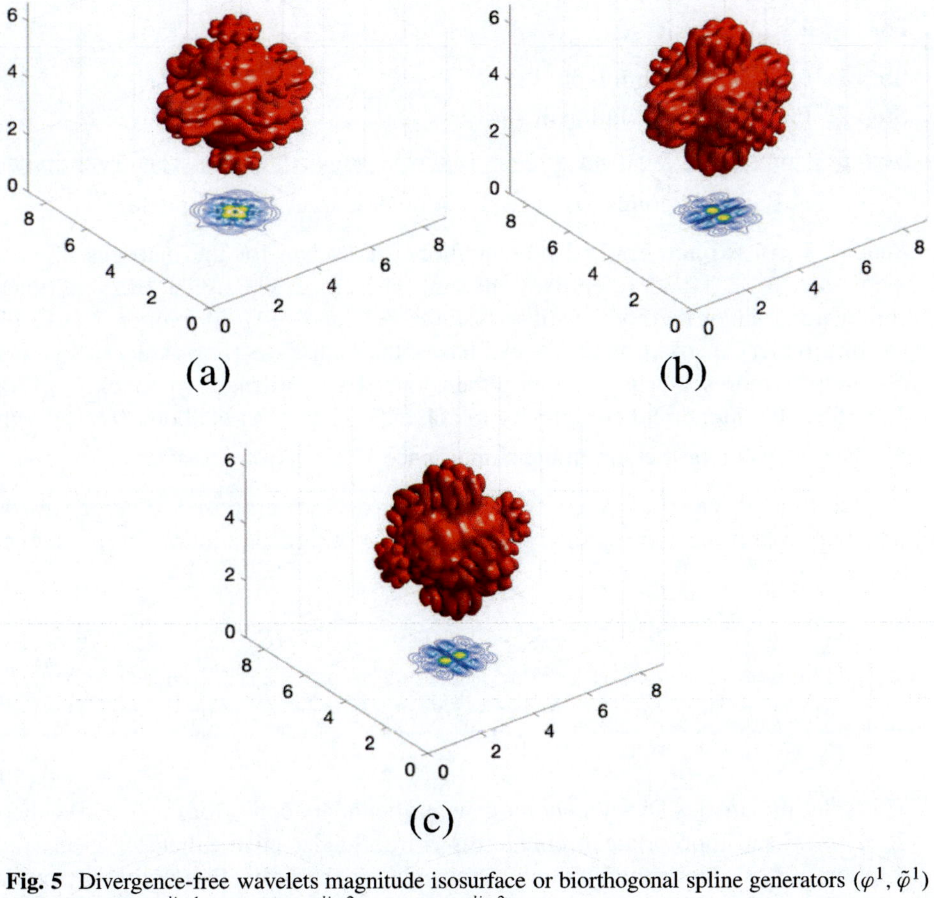

Fig. 5 Divergence-free wavelets magnitude isosurface or biorthogonal spline generators $(\varphi^1, \tilde{\varphi}^1)$ of order 4. **(a)** $|\Psi^{div,1}_{j,k}| = 0.4$. **(b)** $|\Psi^{div,2}_{j,k}| = 0.4$. **(c)** $|\Psi^{div,3}_{j,k}| = 0.4$

The spatial discretization of the divergence-free wavelet-based projection method uses the scaling function and wavelet generators of Sect. 4.1. In practice, one time iteration step (33–34) splits into the following steps:

Step 0: Start with an initial velocity $\boldsymbol{v}^0(x) = \boldsymbol{v}(x, 0)$ defined on dyadic grid points at space resolution $j > j_{jmin}$.

- Compute its wavelet coefficients $[d^{0,\epsilon}_{\boldsymbol{j},\boldsymbol{k}}]$.
- Compute the nonlinear term $(\boldsymbol{v}^0 \cdot \nabla)\boldsymbol{v}^0$ using a fourth-order finite difference scheme, and its wavelet coefficients $[h^{0,\epsilon}_{\boldsymbol{j},\boldsymbol{k}}]$ in $\mathbf{V}^d_j$.

For $1 \le n \le N$, repeat

Step 1: Find $[\tilde{d}^{n+1,\epsilon}_{\boldsymbol{j},\boldsymbol{k}}]$ solution of (33).
Step 2: Find $[d^{div,n+1,\epsilon}_{\boldsymbol{j},\boldsymbol{k}}]$ solution of (34).
Step 3: Compute $[d^{n+1,\epsilon}_{\boldsymbol{j},\boldsymbol{k}}]$ from $[\boldsymbol{d}^{div,n+1,\epsilon}_{\boldsymbol{j},\boldsymbol{k}}]$ using a change of basis [21] and extrapolate $\boldsymbol{v}^{n+1}_j$ at grid points to compute the nonlinear term, and update.

Remark 1 An explicit optimal preconditioner is known for the matrices $\mathbb{M}_j$ and $\mathbb{M}^{div}_j$, see [6, 21] and references therein. Therefore, we use a preconditioned conjugate gradient method to solve systems (33) and (34). Moreover, due to the tensor product construction of wavelet bases, the matrix-vector product in (33) and (34) only requires the use of one-dimensional basis stiffness matrices [20, 22]. Therefore, the numerical complexity to solve (33) and (34) is about $O(N_j^4)$, with N_j the dimension of the one-dimensional space V_j^1.

The time discretization convergence rate of the proposed divergence-free wavelet-based projection method is studied on a designed solution $(\boldsymbol{v}, p)$ defined by

$$\begin{cases} v_1(x, y, z, t) = 2e^{-t}(x^4 + x^2 - 2x^3)(2y + 4y^3 - 6y^2)(2z + 4z^3 - 6z^2), \\ v_2(x, y, z, t) = -e^{-t}(2x + 4x^3 - 6x^2)(y^4 + y^2 - 2y^3)(2z + 4z^3 - 6z^2), \\ v_3(x, y, z, t) = -e^{-t}(2x + 4x^3 - 6x^2)(2y + 4y^3 - 6y^2)(z^4 + z^2 - 2z^3), \\ p(x, y, t) = \cos(t)[(x^2 - x)(y^2 - y)(z^2 - z)]^2. \end{cases} \tag{59}$$

This solution satisfies Dirichlet homogeneous boundary conditions $\boldsymbol{v}_{|\partial\Omega} = 0$, where $\Omega = [0, 1]^3$ and appropriate forcing terms $\boldsymbol{f}$ are added to ensure that (59) is an exact solution of (27). Since the quadrature formula for the projection onto $\mathbf{V}^d_j$ is exact up to order 3, which is the polynomial reproduction order of the scaling function φ^0, the spatial discretization error for the solution (59) is negligible compared to the time discretization error for $\delta t < 0.1$. Table 1 shows different errors between the exact solution projected onto $\mathbf{V}^d_j$ (with a space resolution fixed at $j = 7$ and viscosity $\nu = 2^{-j}$), and the numerical solution of (29–30)–(33–34), in terms of the discretization time step δt. For each norm considered, the expected first-order convergence rate is obtained.

Table 1 Time discretization relative errors according to the time step δt, for the solution (59) at final time $T = 1$, $j = 7$, and $\nu = 2^{-j}$

Backward-Euler				
δt	0.05	0.025	0.0166	Order
L_∞-error	7.2549E^{-5}	3.6427E^{-5}	2.4318E^{-5}	0.99159
L_2-error	2.9129E^{-5}	1.4626E^{-5}	9.7643E^{-6}	0.99157
H^1-error	2.9031E^{-4}	1.4576E^{-4}	9.7312E^{-5}	0.99159

Table 2 Time discretization relative errors according to the time step δt, for the solution (59) at final time $T = 1$, $j = 7$, and $\nu = 2^{-j}$

Crank–Nicolson				
δt	0.05	0.025	0.0166	Order
L_∞-error	6.1014E^{-7}	1.5292E^{-7}	6.8247E^{-8}	1.98770
L_2-error	2.4478E^{-7}	6.1226E^{-7}	2.7271E^{-8}	1.99124
H^1-error	2.4476E^{-6}	6.3177E^{-6}	3.1354E^{-7}	1.87311

Table 3 Spatial discretization errors according to the resolution j, for final time $T = 1$

Backward-Euler				
j	5	6	7	Order
L_2-error	1.40145E^{-3}	6.2469E^{-5}	3.2643E^{-6}	4.3729
L_∞-error	5.4952E^{-3}	2.7106E^{-4}	1.42890E^{-5}	4.2935
H^1-error	9.2027E^{-2}	7.7476E^{-3}	6.5306E^{-4}	3.5693

Going further, we also analyzed the convergence rate of a second-order numerical scheme: Crank–Nicolson for the diffusion part and Adams–Bashforth for the nonlinear term [22]. The results of this experiment are given in Table 2, where again the expected order is achieved.

Similarly, we investigate the spatial projection error convergence rate of the proposed divergence-free wavelet-based projection, using the following exact solution:

$$\begin{cases} v_1(x, y, z, t) = 2e^{-t} \sin^2(2\pi x) \sin(4\pi y) \sin(4\pi z), \\ v_2(x, y, z, t) = -e^{-t} \sin(4\pi x) \sin^2(2\pi y) \sin(4\pi z), \\ v_3(x, y, z, t) = -e^{-t} \sin(4\pi x) \sin(4\pi y) \sin^2(2\pi z), \\ p(x, y, t) = \cos(t)[(x^2 - x)(y^2 - y)(z^2 - z)]^2. \end{cases} \tag{60}$$

The simulation time step is $\delta t = 0.0001$, which is very small compared to the maximal spatial resolution $j = 7$ ($\delta x = 2^{-7} = 0.0078125$) and the used kinematic viscosity $\nu = 2^{-3j}$. Table 3 shows the spatial error for the final time $T = 1$, using the first-order accurate backward-Euler time scheme: as the solution is C^∞, the convergence rate given by Table 3 saturates due to the number of vanishing moments of our wavelet family (equal to 3 in our spline approximation), and we lose one order for the H^1-error. Remark that the use of divergence-free wavelet basis induces no divergence error on the solution $\boldsymbol{v}_j^n$ in $\mathbf{V}_j^d$.

4.3 Simulation of 3D Lid-Driven Flows

In addition to these preliminary studies, the present method was tested on the numerical simulation of 3D lid-driven flows in a cubic cavity $\Omega = [0, 1]^3$, with Reynolds numbers $Re = 100$ and $Re = 1000$. These flows have been extensively studied in the literature, and there are a lot of reference results and solutions, mainly for the two-dimensional flows, see [11]. The simulations used a Matlab code, and the discretization parameters are $\delta t = 0.005$ and $\delta x = 2^{-j}$, where $j = 6$ or $j = 7$ is the considered spatial resolution.

The validation is done by analyzing the flow's steady state. Specifically, we compare our simulation horizontal velocity v_x and vertical velocity v_z profiles in the middle of the cavity to the results of [11]. For $j = 6$, Figs. 6 and 7 show the plot of these profiles, respectively, for the Reynolds numbers $Re = 100$ and $Re = 1000$. As observed, despite of some small discrepancies for $z \in [0, 0.2]$ in the velocity component v_z, our results are in agreement with those of [11]. Moreover, as illustration, we plot in Fig. 8 the velocity magnitude for the simulations performed at resolution $j = 7$. As expected, this highlights the presence of a cavity central vortex.

We also analyze the divergence-free wavelet representation of the solution. For a 3D array $A = [A_{ijk}]$ of scalars, we define its magnitude as $|A| = [|A_{ijk}|]$, the array of coefficients modulus $|A_{ijk}|$. Figures 9 and 10 show the map of divergence-free wavelet coefficient magnitude of the solutions, respectively, for the Reynolds

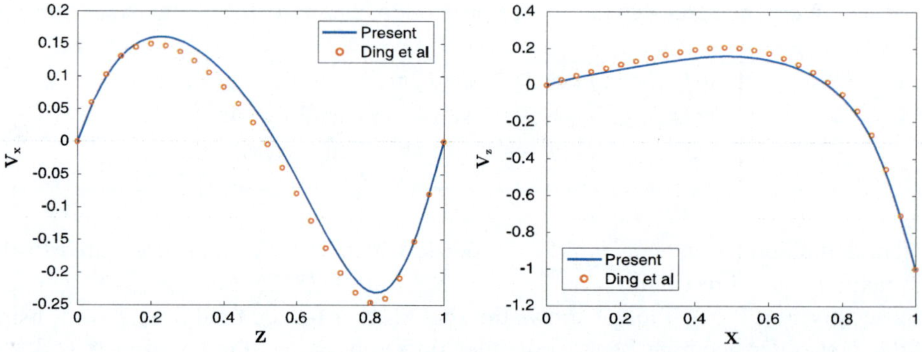

Fig. 6 Steady-state velocity profile in the middle of the cavity: $v_x(0.5, 0.5, z)$ (left) and $v_z(x, 0.5, 0.5)$ (right). Solid line (present work) and circle (reference [11]) for the Reynolds number $Re = 100$ and the spatial resolution $j = 6$

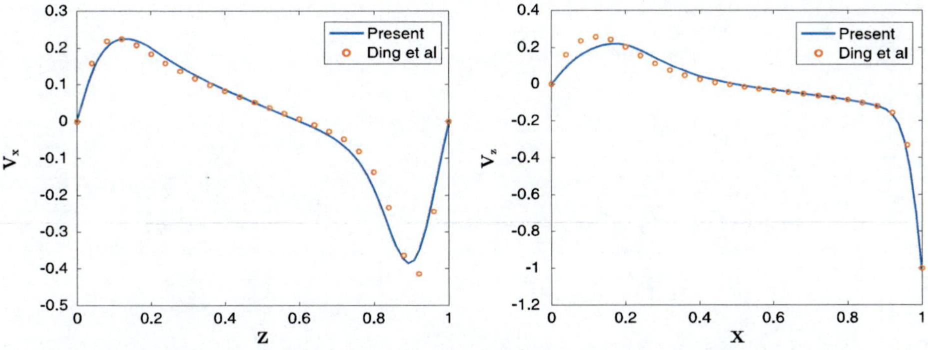

Fig. 7 Steady-state velocity profile in the middle of the cavity: $v_x(0.5, 0.5, z)$ (left) and $v_z(x, 0.5, 0.5)$ (right). Solid line (present work) and circle (reference [11]) for the Reynolds number $Re = 1000$ and the spatial resolution $j = 6$

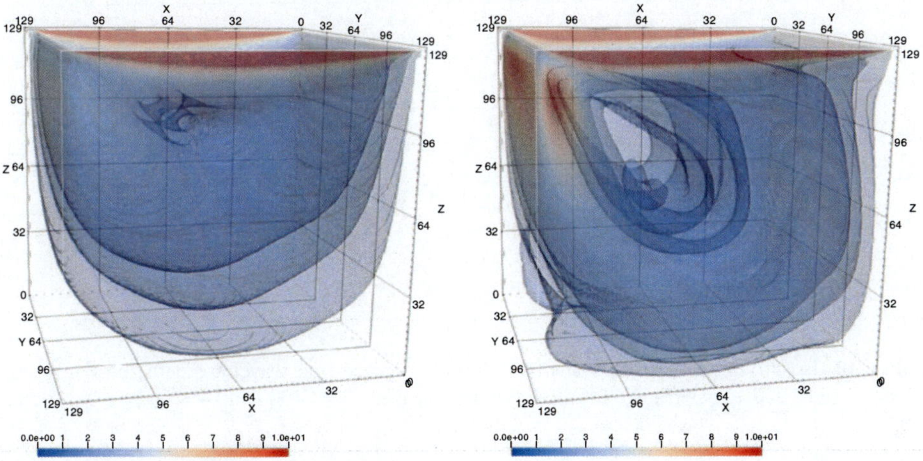

Fig. 8 Isosurface of the steady-state velocity magnitude $|\boldsymbol{v}|$ for the Reynolds number $Re = 100$ (left) and $Re = 1000$ (right), at the spatial resolution $j = 7$

number $Re = 100$ and $Re = 1000$. Clearly, these figures emphasize the quality and the sparsity of such a solution approximation. This suggests the development of adaptive methods with these wavelet bases to improve the numerical complexity. As mentioned before, the actual theoretical complexity of one iteration in the method is about $O(N_j^4)$.

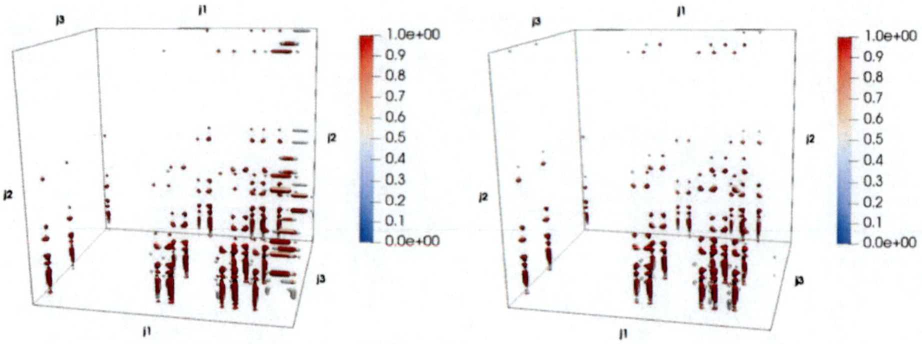

Fig. 9 Isosurface of the steady-state velocity divergence-free wavelets coefficients magnitude $|d^{div,1}_{\mathbf{j},\mathbf{k}}|$ (left) and $|d^{div,2}_{\mathbf{j},\mathbf{k}}|$ (right), for the Reynolds number $Re = 100$ and $4 = j_{min} \leq j_1, j_2, j_3 \leq j = 7$

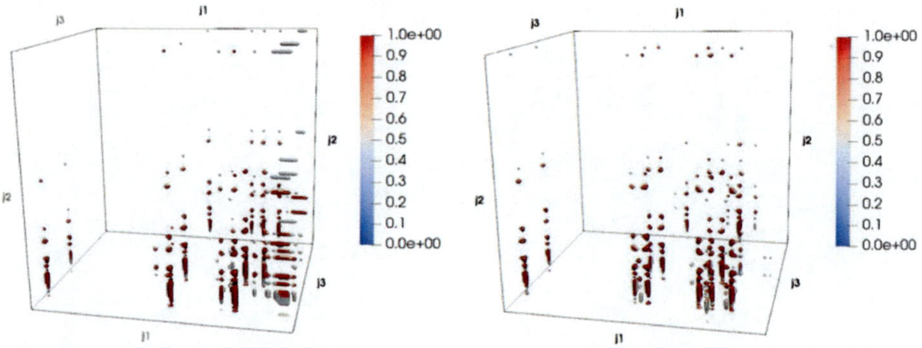

Fig. 10 Isosurface of the steady-state velocity divergence-free wavelets coefficients magnitude $|d^{div,1}_{\mathbf{j},\mathbf{k}}|$ (left) and $|d^{div,2}_{\mathbf{j},\mathbf{k}}|$ (right), for the Reynolds number $Re = 1000$ and $4 = j_{min} \leq j_1, j_2, j_3 \leq j = 7$

5 Conclusion

We presented a construction of wavelets linked by differentiation and integration that allows to construct free-slip and no-slip divergence-free wavelets on the hypercube. These divergence-free wavelets are used to provide the Helmholtz–Hodge decomposition and a change of variables in a first-order accurate time integration for the resolution of Navier–Stokes, similar to the Gauge method. This scheme avoids the use of the projection method Poisson solver that imposes non-physical boundary conditions on the pressure. Our method was tested and validated on the simulation of the well-known 3D lid-driven cavity flow for the moderate Reynolds numbers $Re = 100$ and $Re = 1000$. From the simplicity and precision of this method, we claim that adaptive algorithms can be developed within this approach: mainly for the simulation of turbulent flows at high Reynolds number, including sub-grid models near walls.

Appendix

Lemma 2 *[19, 25] Let $\theta \in L^2(\mathbb{R})$ be a compactly supported function, C^α-differentiable, $\alpha > 0$, and with at least one vanishing moment $\int_{\mathbb{R}} \theta = 0$. Then, there is a positive constant $C(\theta)$ such that*

$$\forall\, (\lambda_{j,k}) \in \ell^2(\mathbb{Z}^2), \quad \| \sum_{j,k\in\mathbb{Z}} \lambda_{j,k}\theta_{j,k} \|_{L^2(\mathbb{R})} \leq C(\theta) \left(\sum_{j,k\in\mathbb{Z}} |\lambda_{j,k}|^2 \right)^{1/2}, \tag{A.1}$$

with $\theta_{j,k} = 2^{j/2}\theta(2^j . - k)$.

Proof. The C^α assumption on θ leads to $|\langle \theta_{j,k}, \theta_{j',k'}\rangle| \leq C 2^{-|j-j'|(1/2+\alpha)}$. As θ is compactly supported, for fixed indices j, k, and j', we have $\langle \theta_{j,k}, \theta_{j',k'}\rangle = 0$ except for some k': $k' \leq M$ if $j' \leq j$ and $k' \leq M 2^{(j'-j)}$ if $j' \geq j$, where M is a positive constant independent of j, j', and k. Then,

$$\sup_{j,k} \sum_{j',k'} |\langle \theta_{j,k}, \theta_{j',k'}\rangle| 2^{(j-j')/2} < +\infty,$$

and since

$$\| \sum_{j,k\in\mathbb{Z}} \lambda_{j,k}\theta_{j,k} \|^2_{L^2(\mathbb{R})} \leq \left(\sum_{j',k'} \sum_{j,k\in\mathbb{Z}} |\lambda_{j,k}|^2 |\langle \theta_{j,k}, \theta_{j',k'}\rangle| 2^{(j-j')/2} \right)^{1/2}$$
$$\cdot \left(\sum_{j',k'} \sum_{j,k\in\mathbb{Z}} |\lambda_{j,'k'}|^2 |\langle \theta_{j,k}, \theta_{j',k'}\rangle| 2^{(j'-j)/2} \right)^{1/2},$$

□

the lemma is proved.

References

1. Beylkin, G.: On the representation of operator in bases of compactly supported wavelets. SIAM J. Numer. Anal. **6**, 1716–1740 (1992)
2. Canuto, C., Masson, R.: Stabilized wavelet approximations of the Stokes problem. Math. Comp. **70**, 1397–1416 (2001)
3. Charton, P., Perrier, V.: A pseudo-wavelet scheme for the two-dimensional Navier-Stokes equations. Comp. Appl. Math. **15**, 137–157 (1996)
4. Chorin, A.J.: Numerical simulation of the Navier-Stokes equation. Math. Comp. **22**, 745–762 (1968)

5. Cohen, A., Daubechies, I., Vial, P.: Wavelets on the interval and fast wavelet transforms. Appl. Comput. Harmon. Anal. **1**, 54–81 (1993)
6. Cohen, A., Masson, R.: Wavelet methods for second order elliptic problems – preconditioning and adaptivity. SIAM J. Sci. Comp. **21**, 1006–1026 (1999)
7. Dahmen, W., Urban, K., Vorloeper, J.: Adaptive wavelet methods-basic concepts and applications to the Stokes problem. In: Zhou, D.-X. (ed.) Wavelet Analysis–Twenty Years Developments, pp. 39–80. World Scientific, New Jersey, (2002)
8. Deriaz, E., Perrier, V.: Divergence-free and curl-free wavelets in 2D and 3D, application to turbulent flows. J. Turbu. **7**, 1–37 (2006)
9. Deriaz, E., Perrier, V.: Direct numerical simulation of turbulence using divergence-free wavelets. SIAM Multis. Model. Simul. **7**, 1101–1129 (2008)
10. Deriaz, E., Perrier, V.: Orthogonal Helmholtz decomposition in arbitrary dimension using divergence-free and curl-free wavelets. Appl. Comput. Harmon. Anal. **26**, 249–269 (2009)
11. Ding, H., Shu, C., Yeo, K.S., Xub, D.: Numerical computation of three-dimensional incompressible viscous flows in the primitive variable form by local multiquadric differential quadrature method. Comput. Methods Appl. Mech. Eng. **195**, 516–533 (2006)
12. Elsasser, W.M.: The hydromagnetic equations. Phy. Rev. **79**, 183–183 (1950)
13. Farge, M.: Wavelet transforms and their applications to turbulence. Ann. Rev. Fluid Mech. **24**, 395–457 (1992)
14. Farge, M., Kevlahan, N., Perrier, V., Goirand, E.: Wavelets and turbulence. Proc. IEEE **84**, 639–669 (1996)
15. Frohlich, J., Schneider, K.: Numerical simulation of decaying turbulence in an adaptive wavelet basis. Appl. Comput. Harm. Anal. **3**, 393–397 (1996)
16. Girault, V., Raviart, P.A.: Finite element methods for Navier-Stokes equations. Springer, Berlin (1986)
17. Grivet-Talocia, S., Tabacco, A.: Wavelets on the interval with optimal localization. Math. Models. Meth. Appl. Sci. **10**, 441–462 (2000)
18. Henry, M., Maitre, E., Perrier, V.: Optimal Transport Using HelmHoltz-Hodge Decomposition and First Order Primal-Dual Algorithm. IEEE ICIP, Piscataway, pp. 4748–4752 (2015)
19. Jouini, A., Lemarié-Rieusset, P.G.: Analyse multirésolution biorthogonale sur l'intervalle et applications. Ann. Inst. Henri Poincaré Sect. C **10**, 453–476 (1993)
20. Kadri-Harouna, S., Perrier, V.: Helmholtz-Hodge decomposition on $[0, 1]^d$ by divergence-free and curl-free wavelets. In: Boissonnat, J.-D., Chenin, P., Cohen, A., Gout, C., Lyche, T., Mazure, M.-L., Schumaker, L. (eds.) Curves and Surfaces, Proceedings of the 7th International Conference, Avignon, June 24–30, 2010. Lecture Notes in Computer Science Series, vol. 6920, pp. 311–329. Springer, Berlin (2012)
21. Kadri-Harouna, S., Perrier, V.: Effective construction of divergence-free wavelets on the square. J. Comput. Appl. Math. **240**, 74–86 (2013)
22. Kadri-Harouna, S., Perrier, V.: Divergence-free wavelet projection method for incompressible viscous flow on the square. Multiscale Model. Simul. **13**, 399–422 (2015)
23. Kadri-Harouna, S., Perrier, V.: Homogeneous Dirichlet wavelets on the interval diagonalizing the derivative operator, and related applications. Preprint hal-01568431v2 (2018)
24. Kim, J., Moin, P.: Application of a fractional-step method to incompressible Navier-Stokes equations. J. Comp. Phys. **59**, 308–323 (1985)
25. Lemarié-Rieusset, P.G.: Analyses multi-résolutions non orthogonales, commutation entre projecteurs et dérivation et ondelettes vecteurs à divergence nulle. Rev. Mat. Iberoamericana **8**, 221–236 (1992)
26. Liu, J.-G., Liu, J., Pego, R.: Stable and accurate pressure approximation for unsteady incompressible viscous flow. J. Comput. Phys. **229**, 3428–3453 (2010)
27. Masson, R.: Biorthogonal spline wavelets on the interval for the resolution of boundary problems. Math. Models Methods Appl. Sci. **6**, 749–791 (1996)
28. Monasse, P., Perrier, V.: Orthogonal wavelet bases adapted for partial differential equations with boundary conditions. SIAM J. Math. Anal. **29**, 1040–1065 (1998)

29. Sass-Hansen, M., Larsen, R., Christensen, N.-V.: Curl-gradient image warping-introducing deformation potentials for medical image registration using Helmholtz decomposition. In: International Conference on Computer Vision Theory and Applications, vol. 1, pp. 79–185 (2009)
30. Schneider, K., Vasilyev, O.: Wavelet methods in computational fluid dynamics. Ann. Rev. Fluid Mech. **42**, 473–503 (2010)
31. Stevenson, R.: Divergence-free wavelet bases on the hypercube: free-slip boundary conditions, and applications for solving the instationary Stokes equations. Math. Comp. **80**, 1499–1523 (2011)
32. Stevenson, R.: Divergence-free wavelets on the hypercube: general boundary conditions. Const. Approx. **44**, 233–267 (2016)
33. Temam, R.: Sur l'approximation de la solution des équations de Navier-Stokes par la méthode des pas fractionnaires II. Arch. Rational Mech. Anal. **33**, 377–385 (1969)
34. Temam, R.: Navier Stokes Equations. North Holland, New York (1977)
35. Urban, K.: Using divergence-free wavelets for the numerical solution of the Stokes problem. In: AMLI'96: Proceedings of the Conference on Algebraic Multilevel Iteration Methods with Applications University of Nijmegen, vol. 2, pp. 261–277 (1996)
36. Urban, K.: Wavelet bases in $H(div)$ and H($curl$). Math. Comput. **70**, 739–766 (2000)
37. Wang, C., Liu, J-G.: Convergence of Gauge method for incompressible flow. Math. Comput. **69**, 1385–1407 (2000)
38. Weinan, E., Guo-Liu , J.: Projection method I: convergence and numerical boundary layers. SIAM J. Numer. Anal. **32**, 1017–1057 (1995)

An Immersed Boundary Method on Cartesian Adaptive Grids for the Simulation of Compressible Flows

S. Péron, T. Renaud, C. Benoit, and I. Mary

Abstract In this article, we present an immersed boundary method (IBM) for the simulation of compressible flows encountered in aerodynamics. The immersed boundary methods allow the mesh not to conform to obstacles, whose influence is taken into account by modifying the governing equations locally (either by a source term within the equation or by imposing the flow variables or fluxes locally, similarly to a boundary condition).

A main feature of the approach we propose is that it relies on structured Cartesian grids in combination with a dedicated HPC Cartesian solver, taking advantage of not only their low memory and CPU time requirements but also the automation of the mesh generation and adaptation. Turbulent flow simulations are performed with Reynolds-Averaged Navier–Stokes equations or with Large-Eddy Simulation approach, in combination with a wall function at high Reynolds number, in order to mitigate the cell count resulting from the isotropic nature of Cartesian cells.

The objective of this paper is to demonstrate the capability of the present immersed boundary method on Cartesian adaptive grids to capture compressible flow features. Results obtained are in good agreement with classical body-fitted approaches but with a significant reduction of the time of the whole process, that is, a day for RANS simulations, including the mesh generation.

1 Introduction

The rise of Computational Fluid Dynamics (CFD) in aerospace sciences in the past decades is due to the growth of the computational power in combination with the increase of robustness and accuracy of CFD solvers. Today, Reynolds-Averaged Navier–Stokes (RANS) simulations on body-fitted meshes are commonly

S. Péron (✉) · T. Renaud · C. Benoit · I. Mary
ONERA, Université Paris Saclay, Châtillon, France
e-mail: stephanie.peron@onera.fr; thomas.renaud@onera.fr; christophe.benoit@onera.fr; ivan.mary@onera.fr

R. Deiterding et al. (eds.), *Cartesian CFD Methods for Complex Applications*, SEMA SIMAI Springer Series 3, https://doi.org/10.1007/978-3-030-61761-5_4

performed by the aeronautical industry in the design phase. The geometrical complexity of the configurations has increased too, taking into account for more details, such as track fairings on an aircraft or rotor head components for a helicopter. Consequently, the mesh generation, which requires usually manual interaction and expertise, has become a major bottleneck of the CFD workflow. This means that efficient tools are required to perform parametric studies and evaluate quickly the impact of a modification of a shape or some details onto the performances of an aircraft. High-fidelity CFD tools are generally not necessary at this stage; lifting-line tools can be used to get trends quickly, but models are often limited to certain flow assumptions. Low-fidelity CFD (e.g., Euler solutions) could be appropriate, but automatic mesh generation is the barrier to override. The immersed boundary methods (IBMs) can be seen as a good compromise between the quality of the solution and how quickly it can be obtained. This concept refers initially to the work of Peskin [32, 33], which employed a novel approach many decades ago to simulate biological flows onto Cartesian grids which did not conform to the geometry. The obstacles lying in the flow are taken into account by introducing a forcing term into the momentum equations. Since then, many variants of this approach have been developed, as quoted by Mittal and Iaccarino [25]. A first approach consists in introducing a continuous source term and is well suited for flows with immersed elastic boundaries [6, 32]. In this context, the source term represents the exchange of momentum between the fluid and solid through a law based on the theory of elasticity. However, in the limit of rigid boundaries, this problem is stiff, leading to a lack of stability and accuracy. Several discrete forcing methods have been developed for flow simulations around solid bodies, among which the ghost-cell direct forcing approach, as developed by Mittal et al. [26], Fadlun et al. [16], and Tseng et al. [41]. The IBM can be used on the whole geometry [29, 41] or locally [27, 43] to capture the potential effects of geometrical details. A similar approach consists in cutting cells that intersect the geometry, which has proven efficient and robust for inviscid flow simulations and low Reynolds flows around complex geometries (see Coirier and Powell [12] and Berger and Aftosmis [4]).

The use of Cartesian grids with local grid refinement in combination with embedded obstacles (either with immersed boundary or with cut-cell methods) seems to be well suited for a high level of automation and computational efficiency [4, 8, 29]. Although the use of adaptive Cartesian grids around arbitrary immersed obstacles is conceptually attractive, the resolution of high Reynolds number flows requires wall models [5, 9] to restrict the number of points within the boundary layer.

This paper proposes an efficient, fast, and robust immersed boundary method on adaptive structured Cartesian grids to perform CFD simulations of compressible flows. The method relies on a second-order accurate finite-volume HPC solver dedicated to Cartesian grids, enabling to deal with a wide range of flow regimes, from subsonic to supersonic flows, for steady RANS simulations or Large-Eddy Simulations (LES). Musker's algebraic wall function [28] is applied within the IBM approach on Cartesian grids in order to solve high Reynolds number flows.

This paper is organized as follows: in Sect. 2, the ghost-cell direct forcing IBM approach used here is described. The way the different immersed boundary conditions are reconstructed at each iteration is detailed. Section 3 describes how this approach is meaningful when applied on Cartesian adaptive grids: an automatic workflow starting from input surfaces describing immersed boundaries has been developed, in combination with a dedicated HPC Cartesian solver, providing results within a short timeline. Section 4 presents two IBM simulations: the first test-case that is considered is the supersonic flow around a blunt body, which is a geometrically simple obstacle but demonstrates the capability of the present method to adapt the mesh during the simulation without any effort while increasing the accuracy of the simulation. The second simulation is a Large-Eddy Simulation of the flow around a three-dimensional high-lift airfoil. Results are compared to experimental data and a reference body-fitted solution.

2 Description of the Immersed Boundary Method

2.1 Governing Equations

The Navier–Stokes equations for a compressible flow can be expressed as follows:

$$\begin{cases} \dfrac{\partial \rho}{\partial t} + \dfrac{\partial}{\partial x_j}(\rho u_j) = 0 \\ \dfrac{\partial \rho u_i}{\partial t} + \dfrac{\partial}{\partial x_j}(\rho u_i u_j) = -\dfrac{\partial p}{\partial x_i} + \dfrac{\partial}{\partial x_j}(\sigma_{ij}) \qquad i = 1, 2, 3 \\ \dfrac{\partial \rho E}{\partial t} + \dfrac{\partial}{\partial x_j}((\rho E + p)\, u_j) = -\dfrac{\partial}{\partial x_j}(Q_j) + \dfrac{\partial}{\partial x_j}(\sigma_{ij} u_i), \end{cases} \tag{1}$$

where ρ denotes the fluid density, u the velocity vector, p the pressure, ρE the total energy per unit mass, σ the viscous stress tensor, and Q the heat flux vector. In our approach, the system (1) is solved for interior cells using a cell-centered finite-volume method of second order of accuracy. W will denote the conservative variables $W = (\rho, \rho u, \rho v, \rho w, \rho E)$ in the following. The Reynolds-Averaged Navier–Stokes (RANS) equations are solved with the Spalart–Allmaras turbulence model [39].

2.2 The Immersed Boundary Method

The immersed boundary method described in this paper relies on a ghost-cell direct forcing formulation, derived from the approaches of Fadlun et al. [16] and Tseng and Ferziger [41]. This approach consists in imposing the flow variables W at some

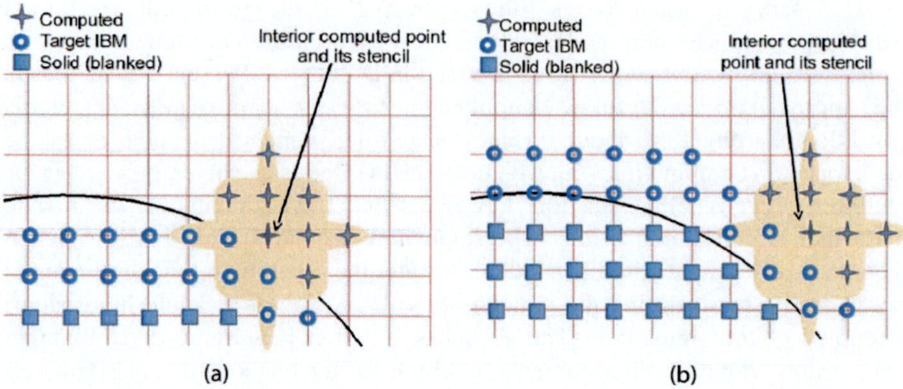

Fig. 1 Spatial stencil involving IB target points for a viscous simulation: with interior (**a**) and exterior (**b**) IB target points. Note that target points can be located either inside or outside the obstacle

particular points, which will be called IB or target points, close to the obstacles to mimic a boundary condition.

First, solid and fluid regions are identified geometrically, by a hole-cutting algorithm [3, 24]. Solid points are marked by squares in Fig. 1. At the fringe of solid region, two layers of IB target points are marked, to be compliant with the numerical scheme, relying on two ghost cells. The solution W is reconstructed at these IB target points using information in the fluid close enough to the wall, at *image points*. Figure 2 displays the case where target IB point A (green dot) is inside the obstacle S. For a sake of simplicity, the image point B (in red dot) can be represented as the symmetrical point of target IB point A with respect to the solid boundary. For that purpose, the distance to the obstacles and also the gradient of the distance to get the normals $\mathbf{n}$ are required. As depicted in Fig. 2, the image points do not usually match fluid points; thus, the solution W at point B is obtained by a second-order interpolation using donor points D_1, D_2, D_3, and D_4. Point N

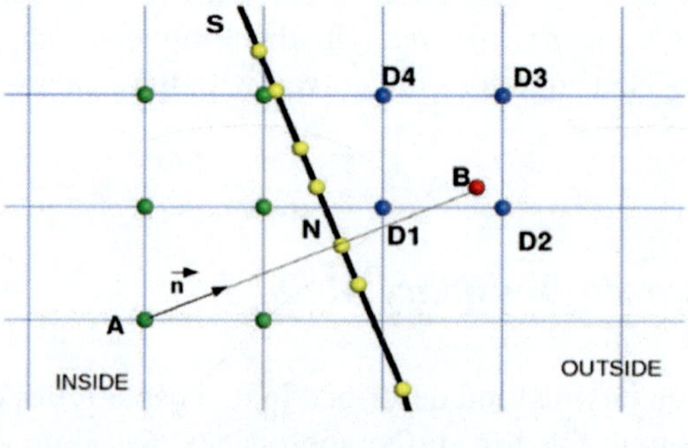

Fig. 2 Sketch describing the present direct forcing IBM approach. Solution W at IB target point A is built up using the corresponding interpolated value of W at its image point B

is the resulting point on the obstacle for which the physical boundary condition shall be recovered implicitly. This point N is obtained by a projection following the normals $\mathbf{n}$.

The update of the solution W at time iteration $n + 1$ using an explicit time integration scheme can be summarized as follows:

- Computation of the flow primitive variables $W = (\rho, u, v, w, T)$ (where ρ is the fluid density, u, v, and w are the velocity vector components in the Cartesian frame, and T is the temperature) for fluid points: $W_{i,j}^{n+1} = f(W_{k,l}^{n})$, (k, l) being the indices of the points in the spatial stencil centered on point of index (i, j)
- Computation of $\hat{W}^{n+1}$ by interpolation of its neighboring values W^{n+1} (superscript $\hat{}$ denotes an interpolated value)
- Reconstruction of the flow variables at IB target points $\tilde{W}^{n+1}$, according to the interpolated value $\hat{W}^{n+1}$.
- Update of ghost cell values at the borders of abutting grids.

2.3 Types of Immersed Boundary Conditions

The reconstruction at IB target points depends on the type of the immersed boundary condition (IBC) defined locally by the input surface.

2.3.1 Wall Slip and No-Slip IBCs

As displayed in Fig. 2, the image point is not necessarily the mirror point of the IB target point with respect to the wall. Thus, slip and no-slip boundary conditions can be implicitly recovered at the wall by a linear reconstruction of the normal component of the velocity, $\mathbf{u_n} = 0$ and $\mathbf{u} = 0$, respectively.

In the case of a no-slip boundary condition where $\|\mathbf{u}\|$=0 at the wall, a one-dimensional linear interpolation is applied given $\mathbf{u}(B)$ as follows:

$$\mathbf{u}(A) = \frac{\Delta_{A,N}}{\Delta_{B,N}} \mathbf{u}(B) ,$$

where $\Delta_{A,N}$ and $\Delta_{B,N}$ are the signed distance of IB target point A and IB image point B to the wall point N, respectively.

In the case of a slip boundary condition, the velocity vector $\mathbf{u}$ can be decomposed in a normal vector and a tangential vector as

$$\mathbf{u} = \mathbf{u}_t + \mathbf{u}_n .$$

The normal velocity vector is obtained by a linear reconstruction, similar to the one applied on $\mathbf{u}$ for the no-slip boundary condition. The tangential velocity vector is then obtained by $\mathbf{u}_t(A) = \mathbf{u}(B) - \|\mathbf{u}_n(B)\|\mathbf{n}$, where $\|\mathbf{u}_n(B)\|$ is the magnitude of the normal velocity at IB image point B and $\mathbf{n}$ is the normal vector to the wall defined at point A.

Pressure and density gradients are assumed equal to zero in the normal direction to the wall in the close vicinity to the wall, and hence $p(A)=p(B)$ and $\rho(A)=\rho(B)$. The pseudo-viscosity $\tilde{\nu}$ of Spalart–Allmaras one-equation turbulence model is recovered by the same linear interpolation such that $\tilde{\nu}$ is implicitly zero at the wall.

2.3.2 Wall Function for High Reynolds Flow Simulations

Our approach relies on an IBM approach on adaptive Cartesian grids, leading to prohibitive cell counts to resolve the viscous stress in the boundary layer until the wall. Moreover, this method is devoted to aeronautical configurations where high Reynolds numbers are often considered. This issue is a key point addressed by many researchers in the field of IBMs, using a wall function to represent the wall shear stress in the case of viscous flow simulations at high Reynolds numbers [5, 9, 10]. In our approach, Musker's algebraic wall function [28] is used to reconstruct the velocity at IB target points, enabling to place first Cartesian cells near the walls at $y^+ \approx 100$. Details on the wall function are provided in [31]. Figure 3 displays the IB target point A and its image point B, for which the variables W_B are interpolated from the computed cells. Instead of a linear reconstruction to recover $u = 0$ at the wall, a wall function is applied between the image point B and the wall. Similarly to the slip boundary condition, the velocity vector is decomposed into a tangential vector and a normal vector. The tangential velocity vector at point A is obtained as follows:

$$\mathbf{u}_t(A) = \frac{U_A}{U_B}\mathbf{u}_t(B)\,,$$

where $U_P = \|\mathbf{u}_t(P)\|$ denotes the magnitude of the tangential velocity at any point P.

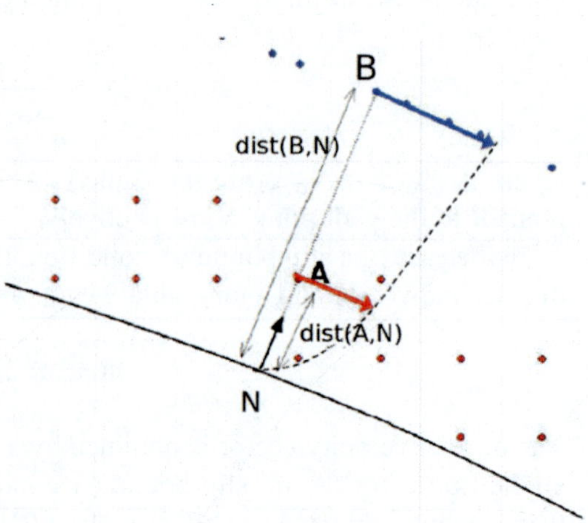

Fig. 3 Wall function for IBM: IB target point is A and corresponding image point is B. In red dots are IB target points around the obstacle; in blue dots, their image points

The velocity vector at point B is obtained by interpolation from its neighboring points. Knowing the modulus of the tangential velocity at image point B, the friction velocity u_τ is obtained by a Newton–Raphson's method on Musker's algebraic wall function. Then, y^+ at point A is computed by $y^+ = \frac{\Delta_{A,N} u_\tau}{\nu}$. The algebraic function (2) provides the modulus of the tangential velocity vector at point A.

$$u^+ = 5.424 \arctan \left[\frac{2y^+ - 8.15}{16.7} \right] + \log \left[\frac{\left(y^+ + 10.6\right)^{9.6}}{\left(y^{+2} - 8.15\, y^+ + 86\right)^2} \right] - 3.52\,. \tag{2}$$

The normal velocity at image point B is obtained by a simple projection:

$$\mathbf{u}_n(B) = (\mathbf{u}(B) \cdot \mathbf{n})\mathbf{n}\,,$$

where $\mathbf{n}$ denotes the unit normal vector at the wall passing through points A and B.

The tangential velocity at IB image point B can be expressed by

$$\mathbf{u}_t(B) = (\mathbf{u}(B) \cdot \mathbf{t})\mathbf{t}\,.$$

We could have imposed the flow to be locally parallel to the wall, which means $\mathbf{u}_n = 0$, but this tends in practice to delay the separation on massively separated flows. Thus, a 1D linear interpolation is performed to compute the normal velocity:

$$\mathbf{u}_n(A) = \frac{\Delta_{A,N}}{\Delta_{B,N}} \mathbf{u}_n(B)\,.$$

The resulting three components of the velocity vector $\mathbf{u}$ at point A are then obtained by summing the corresponding normal and tangential vector components. In the case of a RANS modeling using Spalart–Allmaras model [39], the pseudo-viscosity $\tilde{\nu}$ must also be estimated at IB target point A. Under the assumption of an equilibrium boundary layer, $\tilde{\nu}$ can be defined as

$$\tilde{\nu} = \frac{1}{f_{v1}} \kappa\, u_\tau\, y\,,$$

where f_{v1} is the damping function of Spalart–Allmaras model, which is a nonlinear function of $\tilde{\nu}$, and thus $\tilde{\nu}$ is also obtained by finding explicitly the root of the resulting quartic equation.

2.3.3 Use of Several Types of Immersed Boundary Conditions for a Given Configuration

Several types of immersed boundary conditions can be defined in a single computation, typically to perform a simulation around a model set in a wind tunnel.

In that case, a wall function is applied at the boundary of the model, a slip boundary condition at wind tunnel walls, and an inflow and an outflow boundary conditions at inlet and outlet, respectively. The nature of the immersed boundary condition to be applied is determined by the nature of the input surface on which the IB target point A is projected. If the projection point N lies on a surface tagged as an injection immersed boundary, then the injection immersed boundary condition is flagged for IB point A.

3 IBM on Adaptive Cartesian Grids

3.1 Motivation

Most immersed boundary methods available in the literature rely on adaptive Cartesian grids: Cartesian embedded methods remove the bottleneck of the mesh generation, since the adaptive Cartesian mesh generation can be easily automated even for arbitrary complex geometries. In order to preserve the simplicity of a pure Cartesian approach, the Cartesian mesh is defined down to the wall, relying on the IBM approach to take into account for obstacles. Cartesian cells cannot be refined down to the wall in general (except those cases where the wall is aligned with an axis), so a wall function is mandatory to compute high Reynolds number flows with a reasonable cell count. The strength of the IBM approach on adaptive Cartesian grids used in combination with a Cartesian CFD solver provides an automated and efficient tool for the simulation of flows around complex geometries, provided the IBM preprocessing is robust and fast.

3.2 Automatic IBM Preprocessing for Complex Geometries

3.2.1 Description of the Workflow

The IBM preprocessing can be separated into the following steps:

- The automatic Cartesian mesh generation from a discretized CAD.
- The computation of information required for the IBC reconstruction at each time step of the flow simulation.

First, a Cartesian mesh is generated automatically. This mesh is made of a set of structured uniform grids. The different refinement levels are managed thanks to an octree structure in 3D and quadtree in 2D [30], enabling to prescribe the mesh resolution near each boundary and within the fluid, in order to avoid coarsening in the wake for instance. Ghost cells are explicitly built, such that an overlapping exists between neighboring grids, with a minimum overlap. An example of a Cartesian mesh generated around a 2D profile is displayed in Fig. 4. To generate that case,

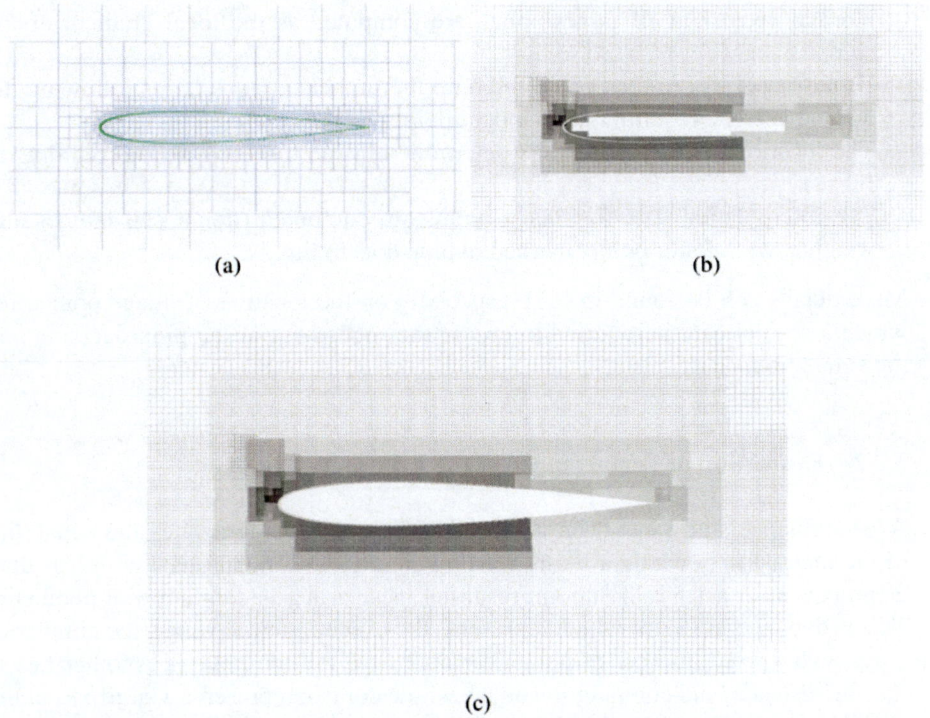

Fig. 4 Example of a mesh around a NACA0012 profile, profile in green. (**a**) Quadtree skeleton mesh in blue around the NACA0012 profile. (**b**) Resulting Cartesian mesh (Cartesian patches fully inside the solid are removed). (**c**) Same Cartesian mesh after blanking of interior cells

the input data are a 1D discretization of the profile and the cell size required in its vicinity (equal to 0.1% of the chord length here). The Cartesian mesh skeleton is a quadtree (i.e., an octree in 2D) mesh, as displayed in Fig. 4a. Each element of the quadtree is then filled with a Cartesian grid of a constant number of cells per direction (specified by the user), resulting in an adaptive Cartesian mesh displayed in Fig. 4b. As shown in this figure, some grids that are entirely inside the solid are removed, to reduce memory requirements. The IBM preprocessing is then achieved, based on several geometrical algorithms initially developed for overset grids [3]. Some of the steps are illustrated by the IBM preprocessing of the previous NACA0012 configuration.

- Interior cells are marked using a blanking technique, either using the X-ray technique introduced by Meakin[24] or using a line-of-sight algorithm [3]. Figure 4b and c represents the same mesh, but blanked-out points are not displayed on the latter.
- The signed distance field is then computed in the whole computational domain (as it is required for the turbulence model later).
- IB target points are marked at the fringe of blanked points (green dots in Fig. 2).

- Normal vectors at IB target points are computed as the local gradient of the signed distance.
- IB target points are then projected onto the immersed boundaries following the normal vectors, resulting in boundary points (yellow dots in Fig. 2).
- The location of image points is determined inside the fluid region (red dots in Fig. 2).
- The interpolation data for image points are computed (donor cell indices and weights, with donor points marked in blue dots in Fig. 2).

More details can be found in [31], especially on the location of image points, for which a special care is required to ensure the robustness of the preprocessing and the simulation.

3.2.2 Evaluation of Performance of the Preprocessing

A performance study has been achieved in [31] that demonstrates the capability of the method to generate a mesh and prepare a CFD simulations within less than 20 min even for a 1.5 billion point resulting mesh around a landing gear geometry. This is depicted in Fig. 5, which represents the execution time versus the number of cores with a fixed number of points per core (five million points). Another fact is that the distance field computation on the whole domain represents a large part of the execution time. Future work will consist in improving this, using a Fast Marching Method [38] for instance, as the distance field must be accurate in the vicinity of the obstacles. As the current distance field computation relies on an orthogonal projection on the triangular surfaces, preconditioned with a k-d tree, further points

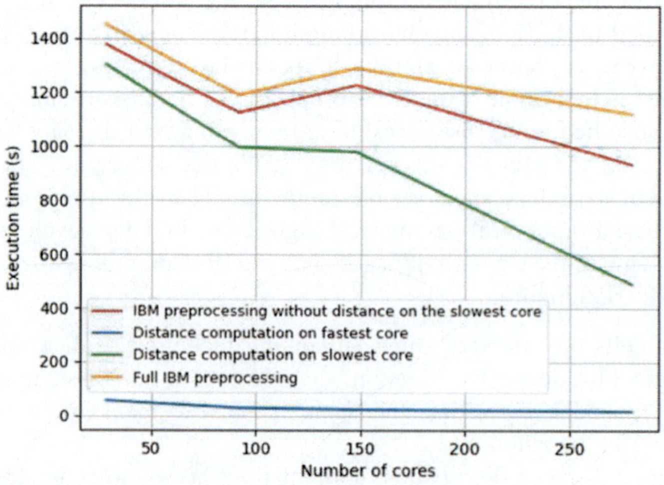

Fig. 5 Weak scaling study: five million nodes distributed per core

are the most time-consuming as the number of candidate triangles for the projection is bigger than for close points to the obstacles.

IBM preprocessing is achieved by an assembly of Cassiopee functions available in several modules (see reference [3] for a general description of Cassiopee or the website [1]).

3.3 IBM Simulations Using a Dedicated Cartesian CFD Solver

3.3.1 FastS HPC Solver

The ONERA HPC FastS solver [2, 22] is used to solve the compressible Navier–Stokes equations using a finite-volume method. It contains a structured multiblock solver that can solve RANS, LES, DNS, steady, and unsteady simulations. Its main feature is its efficiency in dealing with unsteady simulations (see [13]) as it enables to update ten million cells per second per core on a single Intel Broadwell core; in other words, 300 million cells can be updated per second on a 28-core node. FastS contains a solver dedicated to Cartesian grids that is used to perform IBM simulations on Cartesian grids. Despite the relatively high cell count obtained by the block-structured Cartesian mesh generation in comparison with a classical body-fitted unstructured approach, a dedicated Cartesian solver requires much less memory and CPU time than a structured curvilinear solver and also an unstructured solver. Here, the Cartesian solver is 2.5 times more efficient in terms of CPU time and memory than the structured curvilinear solver using the same numerical methods.

FastS solver relies on a hybrid MPI/OpenMP framework, where the memory is distributed (by distributing CFD grids) on the processors at high level, i.e., between nodes, whereas multithreading is managed via OpenMP within a given node. For our purpose, where Cartesian grids are uniform and containing few cells in comparison with grids resolving boundary layers accurately, the N Cartesian grids are distributed between the NT cores using OpenMP.

3.3.2 Numerical Methods

For RANS computations, two spatial schemes are considered, depending on the flow regime: the Roe–MUSCL scheme [36] or an AUSM scheme [23], which is based on a modification of the AUSM+(P) scheme (see Edwards and Liou [15]), which is second-order accurate. Jacobian approximations are those proposed by Jameson and Yoon [18] and Coakley [11], whereas the linear system is solved by the LU-SGS method [18]. The steady and unsteady RANS equations are solved using Spalart–Allmaras one-equation turbulence model [39].

For LES computations, a hybrid centered/upwind scheme [23] is retained to manage a good compromise between robustness and accurate simulation of the turbulent small eddies [19], whereas the temporal integration is achieved by a three-step Runge–Kutta explicit scheme, or by a second-order implicit Gear scheme with local Newton sub-iterations [14].

For Large-Eddy Simulations (LES), the filtered equations are obtained using the formalism developed by Vreman [42]. A Monotone Integrated LES approach (or MILES [7, 17]) is performed, that is, no subgrid-scale model is used as the numerical dissipation of the scheme acts as a subgrid scale model.

3.3.3 Update of IBM Points During the CFD Simulation

The IBM target points must be updated at each sub-step of the time integration. FastS solver updates first fluid cells on each Cartesian grid at time sub-step n, then IB target cells are updated, and finally, transfers between neighboring grids are performed to update the ghost cells. For RANS and LES IBM simulations, Musker's wall model is currently applied at IB target cells only.

MPI transfers between nodes are achieved in a single step: a global transfer to update all the target points and the ghost cells. This is possible because the IBM preprocessing prevents from IB image points to be interpolated by ghost cells (which are explicitly defined in the Cartesian mesh).

In practice, only fluid points are computed by FastS CFD solver, and transfers between abutting grids and IBM updates are performed by a library of Connector module of Cassiopee package [3]. Both FastS and Cassiopee modules handle the same CGNS/Python tree in memory [34, 37]; in other words, arrays defining the CFD simulation (metrics, flow fields) are shared in memory without copy. This is made possible by the fact that ghost cells are explicitly built during the mesh generation, justifying the use of an overset Cartesian mesh, with minimum overlapping.

3.4 Adaptation of the Mesh During the IBM Simulation

The Cartesian mesh adaptation developed for IBM simulations derives from the one developed within an overset grid framework [30], where the near-body regions are defined by a set of structured body-fitted grids close to the obstacles. Similarly, an octree is generated, representing the skeleton of the Cartesian mesh. Finest refinement levels are first defined close to the obstacles, depending on the mesh spacing imposed in the vicinity of them. Similarly to the original approach, where bodies are represented by overset body-fitted grids, the adaptation consists in adapting the skeleton octree according to a refinement indicator, and the set of Cartesian grids is then regenerated. The algorithm is the following:

Step 1 Building of the skeleton octree O_i and generation of the Cartesian set of overlapping grids $\mathcal{M}_i$, where i denotes the adaptation cycle ($i = 0$ initially).

Step 2 IBM preprocessing.

Step 3 N iterations of a CFD simulation (N being big enough to pass the transient phase and to stabilize a mean flow (for unsteady computations) or to converge the solution (for steady computations)).

Step 4 Computation of the physical sensor for adaptation and stored at cell centers. This sensor depends on the physical problem (it can be the vorticity for the wake capture); for unsteady simulations, the maximum value of the sensor during the period is registered, whereas the last value of the sensor is registered for steady simulations.

Step 5 The sensor is then used to compute the indicator on the skeleton octree O_i used to generate the Cartesian mesh O_{i+1} that has been computed. As explained in [30], the number of points after adaptation is controlled (e.g., the increase of the number of points must not exceed 20% of the previous mesh), which provides an automatic values of thresholds of the sensor above and under which octree elements have to be refined or coarsened.

Step 6 Once this indicator is computed, the octree skeleton mesh is adapted, denoted O_{i+1}. A new set of Cartesian grids, denoted $\mathcal{M}_{i+1}$, is then automatically generated.

Step 7 Interpolation of the previous solution defined on mesh $\mathcal{M}_i$ onto the new mesh $\mathcal{M}_{i+1}$.

Step 8 $i = i + 1$, then restart to Step 2 until the criterion that finalizes the adaptation is satisfied. For steady-state simulations, the solution is then converged on the final mesh

For steady flow simulations, the sensor is computed once at convergence and several convergence cycles are performed, whereas for unsteady flows with a stable mean flow, we assume that the mean flow does not vary much during a given period and that the maximum of the sensor for each point is used for the adaptation.

The whole process is automatic, and several adaptations are required to adapt the mesh to the flow features as the number of cells is controlled at each remeshing step and does not exceed 20% more in terms of cell count than the previous mesh. In our approach, the number of adaptation cycles does not exceed generally 5, and no specific criteria based on error estimator for instance are computed to stop the adaptation: as the number of points is increased by 20% at each adaptation (roughly), then 5 adaptations would multiply the number of points by 2.5, which means that during a simulation, one should expect having 2.5 times more computational resources than expected at the first cycle.

Note that the refinement level imposed near each obstacle can be set unchanged during the adaptations even if new refinement levels are created elsewhere.

Adaptation of the octree mesh can also be performed *a priori*: refinement zones can be prescribed in addition to the triangular surfaces describing the obstacles. These refinement zones are closed surfaces. The initial skeleton octree O_0 is then adapted before generating the Cartesian mesh: in that case, the indicator consists in marking octree leaves for refinement if they intersect or lie within the refinement zone and if the cell spacing prescribed for that refinement zone is not reached yet. This approach has been performed on the second test-case in Section 4.2.

4 Numerical Results

A wide range of validations and applications can be found in [35], from Euler to LES simulations, from subsonic to supersonic flows either for two-dimensional academic configurations or for geometrically complex configurations.

Here, we focus on two applications: the first application consists of a simulation of a supersonic flow around a blunt body, in order to demonstrate the capability of the present immersed boundary method to be combined with Cartesian mesh adaptation that occurs periodically during the simulation.

The second test-case is an unsteady three-dimensional simulation around a three-element airfoil, for which the physics is complex. Our objectives are not only to assess first that aerodynamics features on such a case can be captured by a Cartesian method (despite some improvements to be done and further work to be achieved to obtain satisfactory results for aeroacoustics) but also to enhance the HPC capabilities of the whole workflow and especially of the flow solver.

4.1 Validation of the Adaptive Cartesian IBM on a Two-Dimensional Supersonic Case

The bow shock test-case is one of the cases of the International Workshop on High-Order CFD Methods. In this workshop, this case is designed to isolate testing of the shock-capturing properties of schemes using the detached bow shock upstream of a two-dimensional blunt body in supersonic flow. This case is computationally expedient, being steady, two-dimensional, and inviscid, with well-defined boundary conditions. In [21], the author shows that high-order schemes on unstructured grids are able to capture the shock location with very low pressure and enthalpy errors.

The geometry is a flat center section, with two constant radius sections, top and bottom. The flat section is one unit length, and each radius is half the unit length. While the flow is symmetric top and bottom, a full domain is computed to support potentially spurious behavior. The aft section of the body is not included to avoid developing an unsteady wake. The Mach number is $M_\infty = 4$.

Here, our first objective is to validate the IBM approach for supersonic flows. The Cartesian grid is automatically generated around the obstacle. The finest refinement level is imposed not only in the vicinity of the obstacle but also within a refinement zone, which is prescribed prior to the preprocessing. This zone is defined by a circle of radius 3.5 L, where L is the unit length around the blunt body, with a uniform cell size of 0.01 L.

In order to capture the detached shock accurately, an adaptation of the Cartesian mesh is performed periodically during the CFD simulation, as described in Sect. 3.4. The sensor is computed after N iterations (where N is 1000), and it is chosen here as the maximum difference of the Mach number for all the directions of the mesh between the current cell and its direct neighboring cells.

The whole process is automatic and three adaptations are required to adapt the mesh to the shock as the number of cells is controlled at each remeshing step. Figure 6 represents the initial mesh with the refinement zone located at 3.5 times the radius to the center of the blunt body. It is made by 730,000 points over 47 Cartesian grids. The final adapted mesh is made by 1.27 million points over 190 grids, obtained after the three adaptation cycles. In the present adaptation method, new refinement levels can be added; here, the finest refinement level located in the

Fig. 6 Bow shock test-case: Cartesian mesh. (**a**) Initial mesh. (**b**) Adapted mesh

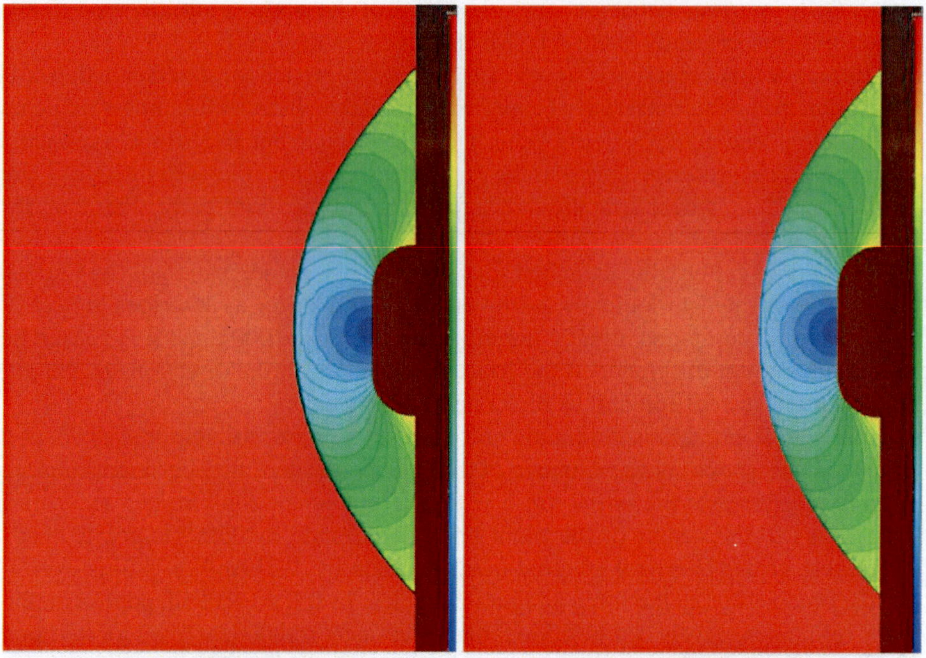

(a) (b)

Fig. 7 Bow shock test-case: isocontours of Mach number. (**a**) Initial mesh. (**b**) Adapted mesh

vicinity of the detached shock is equal to 5.10^{-4} L. It can be noticed that some regions have been coarsened, but the initial spacing near the wall is preserved here.

The CFD simulation has been performed using second-order accurate Roe scheme (with minmod limiter). Figure 7 displays the flow field characteristics through the isocontours of the Mach number: it can be noticed that the shock is better captured after adaptation, which is located around it. Some small oscillations can be noticed near the shock, due to the spatial scheme. A slip immersed boundary condition is applied to model the blunt body.

For a steady inviscid flow, the total enthalpy, $H = \dfrac{\rho E + p}{\rho}$, is constant, where ρE is the total energy, ρ the density, and p the total pressure. The error in this quantity provides a first quantifiable measure of the quality of the computed solution of the general Euler equations (as opposed to schemes that specifically optimize for steady, inviscid flow and enforce H to a constant value). Along the stagnation streamline, the stagnation pressure on the cylinder surface is predicted by the Rayleigh–Pitot formula,

$$\frac{p_{02}}{p_1} = \frac{1 - \gamma + 2\gamma Ma_1^2}{\gamma + 1} \left(\frac{(\gamma + 1)^2 Ma_1^2}{4\gamma Ma_1^2 - 2(\gamma - 1)} \right)^{\frac{\gamma}{\gamma - 1}},$$

Table 1 Comparison of pressure ratios

	Theory	Initial mesh	Adapted mesh
$\frac{p_{02}}{p_1}$	21.068081	21.042294	21.065672
Relative error to theory	0	0.00122398	0.000114344

where p_{02} is the total pressure at stagnation point, p_1 is the static pressure upstream of the shock, Ma is the Mach number, and γ is the specific heat ratio. Subscript $_1$ refers to conditions upstream of the shock and $_2$ to the stagnation point. Table 1 provides a comparison of the wall pressure ratio, showing that the adapted mesh improves this value, by dividing by 10 the relative error with respect to the theory.

4.2 *Unsteady Flow Simulation Around a High-Lift Airfoil*

The test-case is an extruded three-element high-lift airfoil with deployed slat and flap. This kind of configuration is of major interest for acoustics since high-lift devices deployed on aircraft to increase lift at low speed are responsible for a significant part for the airframe noise during the approach phase. An experimental campaign has been conducted in the framework of the joint ONERA/DLR LEISA2 project; experimental data are also provided within the AIAA BANC workshops to validate the numerical methods applied for aerodynamics and acoustics analyses. A reference study is the LES simulation of Terracol and Manoha [40] on a 2.6 billion body-fitted meshes. Six million hours of CPU were required on 4096 processors to perform this simulation. This simulation has also been performed by LBM solvers using an IBM approach on Cartesian grids [20].

The aim of the simulation presented here is to focus only on the aerodynamics phenomena and not on the acoustics, since the way the information is transferred from a grid of a level l to a grid of a different level (twice as coarse) leads to small perturbations that are a major issue for a far-field acoustics analysis.

4.2.1 Description of the Test-Case

The reduced geometry configuration is used here (F16). The retracted wing chord length c is 300 mm. The slat and flap are deployed, respectively, of 27.834° and 35,011°. The flow conditions are $M_\infty = 0.178$, $\alpha = 6.15°$ and a Reynolds number of $Re = 1.23$ million, based on the chord. The wing span is chosen the same as the reference CFD study, that is, 0.25 c.

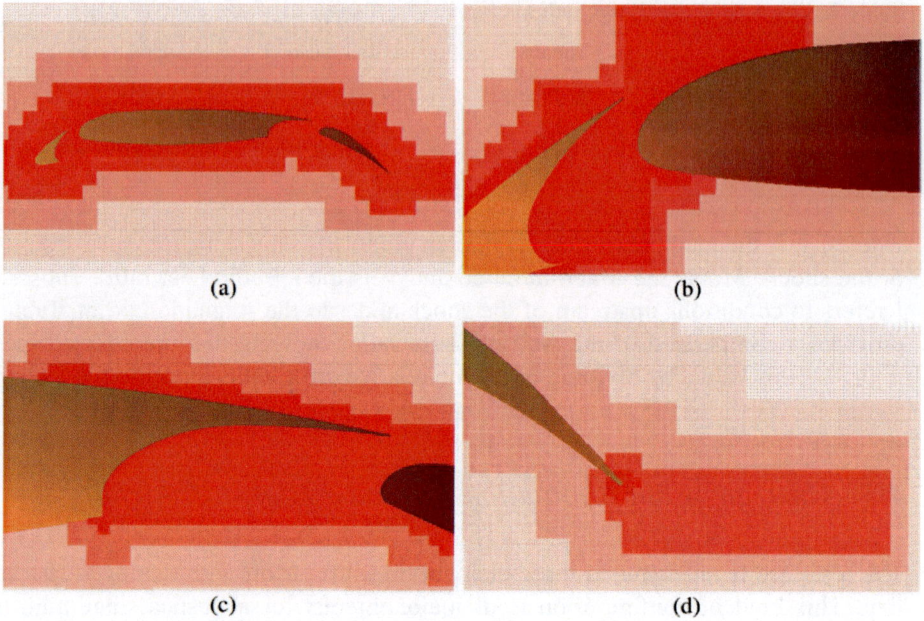

Fig. 8 Views of the IBM Cartesian mesh around the high-lift airfoil. (**a**) General view around the three-element airfoil. (**b**) Close-up view near the slat cove. (**c**) Close-up view near the trailing edge of the main element. (**d**) Close-up view near the trailing edge of the flap

An IBM simulation with FastS solver is performed on a set of Cartesian grids using Musker's wall function applied at IB target points. The mesh is composed by 660 million points, with an adapted spatial resolution in the vicinity of the flap and the slat and in their cavity and wake regions, with a smallest cell size equal to $1.5\ 10^{-4}\ c$. The external border of the computational domain is located at 50 c. The mesh is represented on different views displayed in Fig. 8.

The LES simulation has been initialized by a RANS simulation in order to get rid of transient phenomena. The spatial scheme is the modified AUSM scheme [23], to manage a good compromise between robustness and accurate simulation of the turbulent small eddies [19], whereas the temporal integration is an explicit three-step Runge–Kutta scheme, with a global time step, $\Delta t = 0{,}16\,\mu\text{s}$. The simulation has been performed on 224 Intel Broadwell cores of ONERA SATOR cluster, for a CPU cost of 0.4 μs per point per iteration. The flow quantities have been averaged on a period of 80 ms.

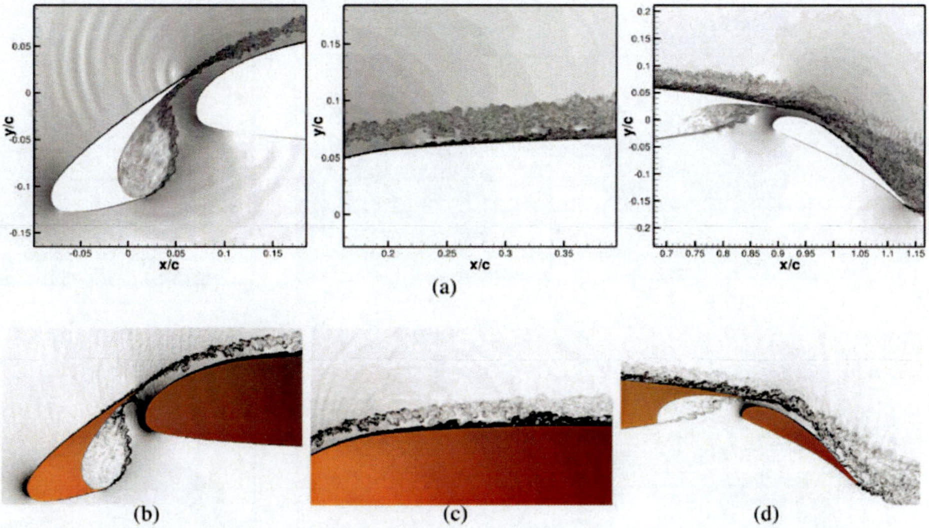

Fig. 9 Instant views of the flow represented by the density gradient: comparison between the wall-resolved LES (**a**) and the IBM simulations (**b**, **c**, and **d**)

4.2.2 Results

Figure 9 displays the density gradient resulting from the LES simulation using the wall-modeled IBM. The comparison with the reference simulation of Terracol and Manoha shows that the IBM approach enables to capture the main features of this flow: recirculation bubble in slat and flap cavities, turbulent boundary layers, and wakes. This is also assessed by the comparison of the averaged velocity between the reference LES and the IBM simulation and experimental data, displayed in Fig. 10. The location of recirculation bubble in cavities is well captured. Besides, the simulated flows in the vicinity of the suction side of the flap differ from the experiments, where a strong separation occurs unlike the LES simulations. Other wind tunnel tests did not reveal that separation, and Terracol [40] dcmonstrated that this difference was due to the influence of the wind tunnel walls.

Two rakes of probes are defined in the fluid, as displayed in Fig. 11. At these locations, the velocity and velocity fluctuation profiles are compared against the experiment and the reference LES body-fitted simulation, as displayed in Fig. 12, showing a good agreement between both simulations and also with the experimental results.

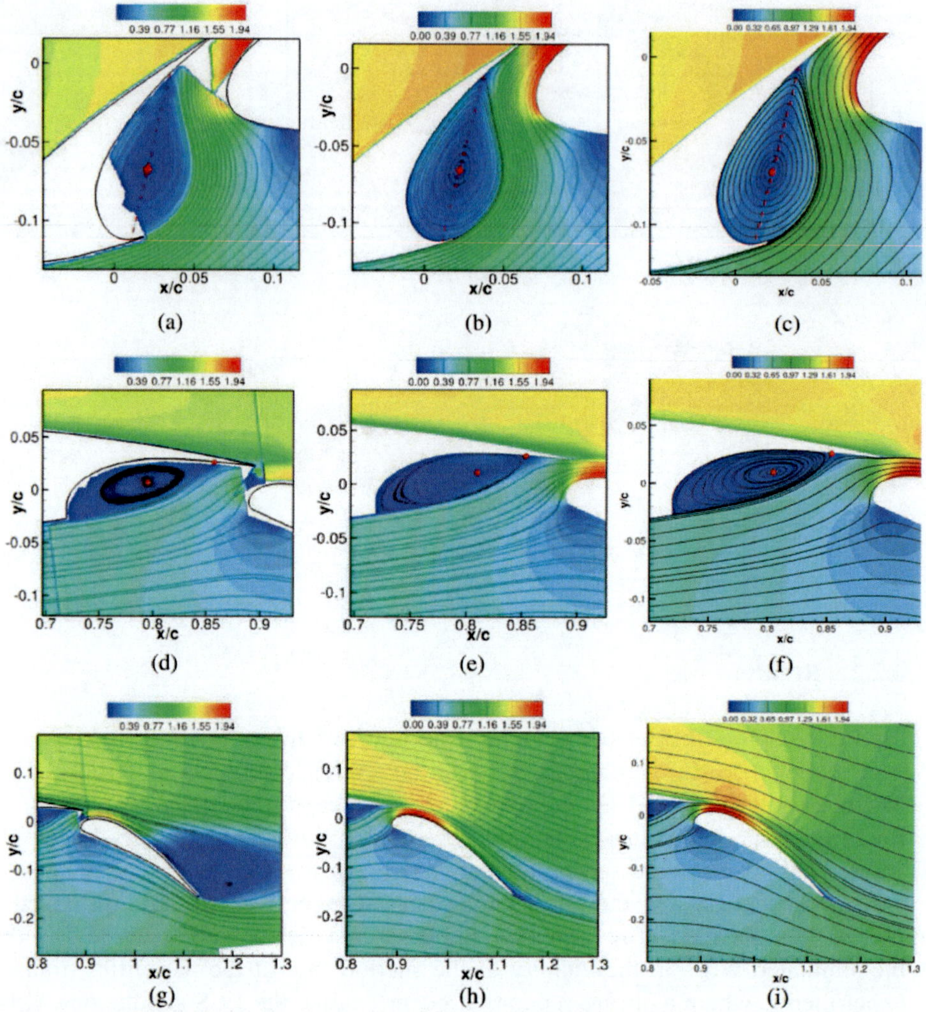

Fig. 10 Views of the averaged flow: isocontours of the velocity amplitude and streamlines; comparison between experiments (left-hand side), the reference wall-resolved LES simulation (center), and the IBM LES simulation (right-hand side). (**a**) Experiments: slat cove. (**b**) Body-fitted LES: slat cove. (**c**) Wall-modeled IBM: slat cove. (**d**) Experiments: main element trailing edge. (**e**) Body-fitted LES: main element trailing edge. (**f**) Wall-modeled IBM: main element trailing edge. (**g**) Experiments: flap. (**h**) Body-fitted LES: flap. (**i**) Wall-modeled IBM: flap

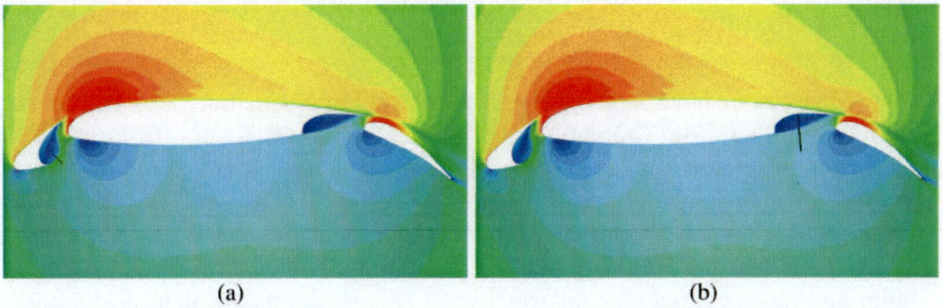

Fig. 11 Probe locations: (**a**) 04-2 in the slat cove and (**b**) 18-3 in the trailing edge cove of the main wing

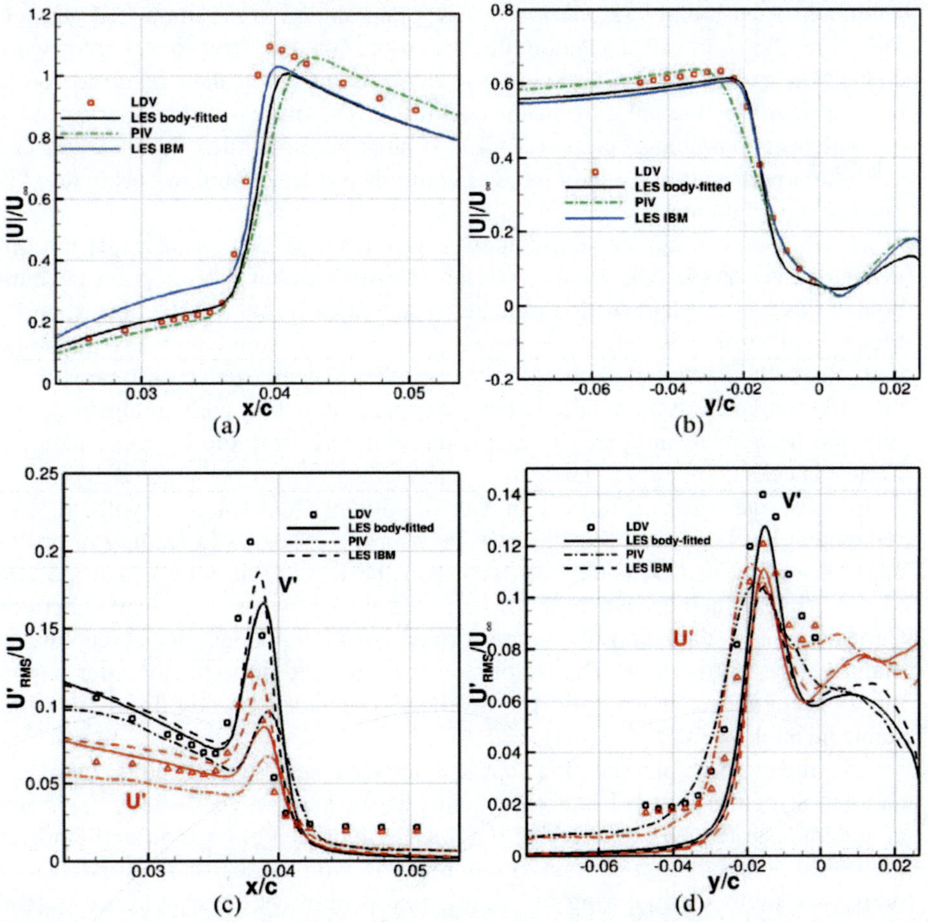

Fig. 12 Comparison of velocity profiles at probe 04-2 (**a**) and at probe 18-3 (**b**); velocity fluctuations at probe 04-2 (**c**) and 18-3 (**d**). IBM simulation is compared against the reference LES simulation, PIV, and LDV data

5 Conclusions

In this chapter, we have presented an immersed boundary method (IBM) on adaptive Cartesian grids for the simulation of compressible flows. As the IBM enables the mesh not to conform to the obstacles, Cartesian meshes are attractive for that purpose as their generation and adaptation are usually fast and easy to perform. Using both approaches together enables to get rid of the mesh generation, which is a tedious task for engineers as the configurations that are simulated become more and more complex geometrically. CFD simulations are then performed by a dedicated HPC Cartesian solver, taking advantage of the low memory and CPU time requirements for Cartesian grids (since metrics do not need to be stored, flux balances are simplified).

The approach that has been developed relies on a workflow that needs to be automatic, robust, and fast, starting from a triangulation of the immersed obstacles only. For that purpose, an automatic preprocessing has been developed, which enables to generate the Cartesian mesh and compute the data required for the reconstruction of the solution in the vicinity of the immersed boundaries. As an example, it is possible to prepare a CFD simulation of an IBM Cartesian mesh of 1.5 billion points around a landing gear within less than 20 min, requiring less than 360 GBytes of memory.

Several types of immersed boundaries have been developed such that not only inviscid or viscous wall boundaries can be reconstructed but also injection and outlet boundaries can be defined as immersed boundaries, provided the corresponding triangulated surface is defined as input. Turbulent flow simulations are performed with Reynolds-Averaged Navier–Stokes equations using Spalart–Allmaras model or with a Large-Eddy Simulation approach, in combination with an algebraic wall function in order to mitigate the cell count resulting from the isotropic nature of Cartesian cells.

Initially, the mesh is refined in the vicinity of the obstacles, with different refinement levels possible for different parts of the obstacles (a leading or trailing edge of a wing can be better resolved than the rest of the wing). It is possible to prescribe refinement zones *a priori* if the flow physics is known; otherwise, Cartesian mesh adaptation can be performed during the simulation to capture the main features of the flow. The adaptation is performed periodically, after passing the transient phase, and is valid for steady flows or for unsteady flows but with a stable mean flow.

To validate that approach, the first test-case has been considered, which is the simulation of the inviscid and supersonic flow around a blunt body. This test-case demonstrates the capability of the present approach to perform automatically immersed boundary simulations in combination with Cartesian mesh adaptation, performed periodically during the simulation to improve the capture of the flow characteristics without knowing it *a priori*.

The second application is an unsteady simulation of the flow around a high-lift airfoil. For that case, refinement zones are prescribed in regions where the wakes are developed. Aerodynamics results are evaluated here and compared with experiments and a reference LES simulation on a structured body-fitted mesh by Terracol [40]. Some works have to be achieved to be able to evaluate the acoustics, as no specific treatment is achieved yet when crossing an interface from a fine grid to a coarser grid (twice as coarse here), leading to reflections of unsupported structures back into the finer grid. This is one topic on which we will focus in the future.

Future work will also concern the improvement of the wall modeling using wall functions, especially to improve the skin friction, which is a major concern for compressible aerodynamics applications, especially for aircrafts. Adaptation of the mesh within the boundary layer is a topic of interest, as the cell spacing close to the wall is currently determined by the user, usually considering the Reynolds number and a y^+ value of 100 roughly computed for a flat plate, which is not really relevant for any configuration.

Adapted wall models for Large-Eddy Simulations will also be considered. Another topic is to extend the method to bodies in relative motion, aiming at simulating flows around configurations such as control surfaces on wings or VTOLs.

References

1. http://elsa.onera.fr/Cassiopee/Userguide.html
2. https://w3.onera.fr/FAST
3. Benoit, C., Péron, S., Landier, S.: Cassiopee: a CFD pre- and post-processing tool. Aerospace Sci. Technol. **45**, 272–283 (2015)
4. Berger, M.J., Aftosmis, M.J.: Progress towards a Cartesian cut-cell method for viscous compressible flow. In: 50th AIAA Aerospace Sciences Meeting Including the New Horizons Forum and Aerospace Exposition, pp. 2012–1301 (2012)
5. Berger, M.J., Aftosmis, M.J.: An ODE-based wall model for turbulent flow simulations. AIAA J., 1–15 (2017)
6. Beyer, R.P., LeVeque, R.J.: Analysis of a one-dimensional model for the immersed boundary method. SIAM J. Numer. Anal. **29**(2), 332–364 (1992)
7. Boris, J.P., Grinstein, F.F., Oran, E.S., Kolbe, R.L.: New insights into large eddy simulation. Fluid Dyn. Res. **10**(4-6), 199–228 (1992)
8. Brehm, C., Barad, M.F., Kiris, C.C.: Open rotor computational aeroacoustic analysis with an immersed boundary method. In: 54th AIAA Aerospace Sciences Meeting, p. 0815 (2016)
9. Capizzano, F.: Turbulent wall model for immersed boundary methods. AIAA J. **49**(11), 2367–2381 (2011)
10. Chen, Z.L., Hickel, S., Devesa, A., Berland, J., Adams, N.A.: Wall modeling for implicit large-eddy simulation and immersed-interface methods. Theor. Comput. Fluid Dyn. **28**(1), 1–21 (2014)
11. Coakley, T.J.: Implicit upwind methods for the compressible Navier-Stokes equations. AIAA J. **23**(3), 374–380 (1985)
12. Coirier, W.J., Powell, K.G.: Solution-adaptive Cartesian cell approach for viscous and inviscid flows. AIAA J. **34**(5), 938–945 (1996)
13. Dandois, J., Mary, I., Brion, V.: Large-eddy simulation of laminar transonic buffet. J. Fluid Mech. **850**, 156–178 (2018)
14. Daude, F., Mary, I., Comte, P.: Self-adaptive Newton-based iteration strategy for the les of turbulent multi-scale flows. Comput. Fluid. **100**, 278–290 (2014)

15. Edwards, J.R., Liou, M.-S.: Low-diffusion flux-splitting methods for flows at all speeds. AIAA J. **36**(9), 1610–1617 (1998)
16. Fadlun, E.A., Verzicco, R., Orlandi, P., Mohd-Yusof, J.: Combined immersed-boundary finite-difference methods for three-dimensional complex flow simulations. J. Comput. Phys. **161**(1), 35–60 (2000)
17. Garnier, E., Mossi, M., Sagaut, P., Comte, P., Deville, M.: On the use of shock-capturing schemes for large-eddy simulation. J. Comput. Phys. **153**(2), 273–311 (1999)
18. Jameson, A., Yoon, S.: Lower-upper implicit schemes with multiple grids for the Euler equations. AIAA J. **25**(7), 929–935 (1987)
19. Laurent, C., Mary, I., Gleize, V., Lerat, A., Arnal, D.: DNS database of a transitional separation bubble on a flat plate and application to RANS modeling validation. Comput. Fluids **61**, 21–30 (2012)
20. Le Garrec, T., Mincu, D.C., Terracol, M., Casalino, D., Ribeiro, A.: Aeroacoustic prediction of the LEISA2 high-lift airfoil: Lattice Boltzmann method vs. Navier-Stokes Finite Volume method and experiments. In: Turbulence and Interactions Conference (2015)
21. Le Gouez, J.M.: A finite volume method for high Mach number flows on high-order grids. In: 7th European Conference on Computational Fluid Dynamics (ECFD 7) (2018)
22. Mary, I.: Flexible Aerodynamic Solver Technology in an HPC environment. Maison de la Simulation Seminars (2016). http://www.maisondelasimulation.fr/seminar/data/201611_slides_1.ppt
23. Mary, I., Sagaut, P.: Large Eddy simulation of flow around an airfoil near stall. AIAA J. **40**(6), 1139–1145 (2002)
24. Meakin, R.L.: Object X-Rays for cutting holes in composite overset structured grids. In: 15th AIAA Computational Fluid Dynamics Conference, pp. 2001–2537 (2001)
25. Mittal, R., Iaccarino, G.: Immersed boundary methods. Ann. Rev. Fluid Mech. **37**, 239–261 (2005)
26. Mittal, R., Dong, H., Bozkurttas, M., Najjar, F.M., Vargas, A., von Loebbecke, A.: A versatile sharp interface immersed boundary method for incompressible flows with complex boundaries. J. Comput. Phys. **227**(10), 4825–4852 (2008)
27. Mochel, L., Weiss, P.-E., Deck, S.: Zonal immersed boundary conditions: application to a high-Reynolds-number afterbody flow. AIAA J. **52**(12), 2782–2794 (2014)
28. Musker, A.J.: Explicit expression for the smooth wall velocity distribution in a turbulent boundary layer. AIAA J. **17**(6), 655–657 (1979)
29. Nakahashi, K.: Immersed boundary method for compressible Euler equations in the Building-Cube Method. AIAA Paper, pp. 2011–3386 (2011)
30. Péron, S., Benoit, C.: Automatic off-body overset adaptive Cartesian mesh method based on an octree approach. J. Comput. Phys. **232**(1), 153–173 (2013)
31. Péron, S., Benoit, C., Renaud, T., Mary, I.: An immersed boundary method on Cartesian adaptive grids for the simulation of compressible flows around arbitrary geometries. Eng. Comput. 1–19 (2020)
32. Peskin, C.S.: Flow patterns around heart valves: a numerical method. J. Comput. Phys. **10**(2), 252–271 (1972)
33. Peskin, C.S.: The immersed boundary method. Acta Numer. **11**, 479–517 (2002)
34. Poinot, M.: Five good reasons to use the hierarchical data format. Comput. Sci. Eng. **12**(5), 84–90 (2010)
35. Renaud, T., Benoit, C., Péron, S., Mary, I., Alferez, N.: Validation of an immersed boundary method for compressible flows. In: AIAA Scitech 2019 Forum. AIAA Paper, pp. 2019–2179 (2019)
36. Roe, P.L.: Approximate Riemann solvers, parameter vectors, and difference schemes. J. Comput. Phys. **43**(2), 357–372 (1981)
37. Rumsey, C.L., Wedan, B., Hauser, T., Poinot, M.: Recent updates to the CFD general notation system (CGNS). In: 50th AIAA Aerospace Sciences Meeting, vol. 10, pp. 6–2012 (2012)
38. Sethian, J.A.: Fast marching methods. SIAM Rev. **41**(2), 199–235 (1999)

39. Spalart, P.R., Allmaras, S.R.: A one-equation turbulence model for aerodynamic flows. AIAA J. 94 (1992)
40. Terracol, M., Manoha, E.: Wall-resolved large eddy simulation of a high-lift airfoil: detailed flow analysis and noise generation study. In: 20th AIAA/CEAS Aeroacoustics Conference. AIAA Paper, pp. 2014-3050 (2014)
41. Tseng, Y.-H., Ferziger, J.H.: A ghost-cell immersed boundary method for flow in complex geometry. J. Comput. Phys. **192**(2), 593–623 (2003)
42. Vreman, A.W.: Direct and Large-Eddy Simulation of the Compressible Turbulent Mixing Layer. Universiteit Twente, Enschede (1995)
43. Zhu, W.J., Behrens, T., Shen, W.Z., Sørensen, J.N.: Hybrid immersed boundary method for airfoils with a trailing-edge flap. AIAA J. **51**(1), 30–41 (2012)

Magnetohydrodynamics Adaptive Solvers in the AMROC Framework for Space Plasma Applications

Müller Moreira Lopes, Margarete Oliveira Domingues, Ralf Deiterding, and Odim Mendes

Abstract Plasma disturbances affect satellites and spacecraft and can cause serious problems to telecommunications and sensitive sensor systems on Earth. Considering the huge scale of the plasma phenomena, data collection at individual locations is not sufficient to cover this entire relevant environment. Therefore, computational plasma modelling has become a significant issue for space sciences, particularly for the near-Earth magnetosphere. However, the simulations of these disturbances present many physical as well as numerical and computational challenges. In this work, we discuss our recent magnetohydrodynamic solver, realised within the MPI-parallel AMROC (Adaptive Mesh Refinement in Object-oriented C++) framework, in which particular physical models and automatic mesh generation procedures have been implemented. A performance analysis using a selection of significant space applications validates the solvers capabilities and confirms the technical importance of our approach.

1 Introduction

The number of studies on space plasmas has increased during the last decades, driven by the significant effects that space plasma can have on sensitive electro-electronic technologies. The road map for 2015–2025 commissioned by the Committee on Space Research (COSPAR) and the International Living With a Star

M. M. Lopes (✉)
National Institute for Space Research (INPE), São José dos Campos, São Paulo, Brazil
e-mail: muller.lopes@inpe.br

M. O. Domingues · O. Mendes
National Institute for Space Research (INPE), São José dos Campos, Brazil
e-mail: margarete.domingues@inpe.br; odim.mendes@inpe.br

R. Deiterding
Aerodynamics and Flight Mechanics Research Group, University of Southampton, Southampton, UK
e-mail: r.deiterding@soton.ac.uk

R. Deiterding et al. (eds.), *Cartesian CFD Methods for Complex Applications*,
SEMA SIMAI Springer Series 3, https://doi.org/10.1007/978-3-030-61761-5_5

working group (ILWS) [43] describes damaging influences of space phenomena on current technologies and infrastructure, implying a high economic cost that—albeit not completely measurable—has a significant impact on the world economy [21].

There are significant international efforts to produce space weather forecasting systems in order to anticipate when very intense solar events can occur and how they can interact with the Earth or human-built space equipment. The global simulations of the magnetohydrodynamics (MHD) model are a fundamental part of such forecasting systems, especially to clarify processes, quantify, and even in the next years predict a complete phenomenology of the interaction of plasmas with the Earth's magnetised and ionised atmosphere, as described, for instance, in [47].

In recent years, our research group has been contributing to such efforts by developing a high performance numerical MHD solver to simulate the near-Earth environment using the solar wind data as boundary conditions. This solver is being developed under the AMROC (Adaptive Mesh Refinement in Object-oriented C++) framework [17], which implements a patch-Structured Adaptive Mesh Refinement (SAMR) method [4] with the parallel strategy proposed in [14] using the Message Passing Interface (MPI) protocol.

Snapshots of these developments are given in [36], which introduces our solver by presenting the results of classical ideal MHD benchmarks using the adaptive and parallel strategies for both two-dimensional and three-dimensional formulations, and in [19], where we presented some improvements concerning the adaptive criteria of the SAMR method by using wavelet-based techniques. A core finding of these publications is that the multiresolution (MR) approach, as it is mathematically more rigorous, leads to an adaptive mesh well fitted to the structures of solutions in the MHD context and to an improvement in the overall computational time. Furthermore, this approach is more suitable for the numerical approaches to solve the magnetic field divergence problem presented in MHD simulations.

In recent years, the number of studies in computational plasma modelling has been increasing; many other MHD codes were developed for a variety of applications. For instance, [22] presents an extensive review of global magnetosphere models, codes and numerical methods. In particular, the RAMSES code [23] was developed in Saclay to study large-scale structure and galaxy formation; however, it is now a rather flexible package to be used for general purpose simulations in self-gravitating fluid dynamics. It is written in Fortran 90 with extensive use of the MPI library and built on a grid-based hydro solver with adaptive mesh refinement. The underlying data structure is the so-called fully threaded tree. As opposed to patch-based SAMR, cells are refined on a cell-by-cell basis; it is therefore called a tree-based AMR as described in [23]. A comparison of advantages and disadvantages of patch-based and tree-based AMR is presented in [24], stating that the patch-based SAMR codes provide a better memory layout and a simpler geometry at the cost of the refinement of unnecessary cells and the extra memory consumption due to the patch manager. Another widely used MHD solver is the ATHENA code [44]. It is a grid-based code for astrophysical magnetohydrodynamics. It was developed primarily for studies of the interstellar medium, star formation and accretion flows. Athena has been made freely available to the community in the hope that others may find it useful. It uses Static Mesh Refinement (SMR), as described in [44]. This approach consists of using a fixed mesh that is more refined in some regions of the domain. Besides being faster, this approach is not as flexible as the full SAMR algorithm in the sense that, for many problems, the structures of interest may be formed dynamically or transported through the domain.

This paper aims to present the new milestones reached in the development of our MHD solver. In particular, this work develops the resistive formulation of the MHD equations using a set of physical benchmark problems that occur due to the resistive effects. Furthermore, we introduce an early two-dimensional MHD formulation for the Earth magnetosphere based on [38–40]. After the simulated magnetosphere converges to steady-state conditions, the model is implemented such that the boundary condition containing typical values for the solar wind proceeds with an evolution using actual interplanetary measurements, for instance, data collected from the OMNI web service, NASA.[1]

The outline of this work briefly presents the AMROC framework in Sect. 2, explaining the numerical, computational and design decisions of the software development process that are contributing to the success of our ideal MHD solver, as discussed in [19, 36]. Then, the numerical and implementation aspects of the ideal and resistive MHD solvers are presented in Sect. 3. In Sect. 4, the content refers to some fundamental experiments related to physical phenomena that occur in space weather. At last, in Sect. 5, we discuss the lessons learned and the next development steps.

2 AMROC

The compressible MHD equations are a system of nonlinear hyperbolic partial differential equations. Considering the vector of conservative physical quantities $\mathbf{q}$, these equations can be written as a conservation law, i.e.

$$\frac{\partial \mathbf{q}}{\partial t} + \nabla \cdot \mathbb{F}(\mathbf{q}) = \mathbf{S}(\mathbf{q}), \tag{1}$$

where $\mathbb{F}$ and $\mathbf{S}$ are the flux function and the source term, respectively. A suitable approach to simulate this type of model is the finite volume scheme [33]. This method consists of discretising the physical domain of the problem into cells or control volumes, so that each cell contains an average value for its coverage. Then, each cell average is evolved in time according to the flux between its adjacent cells. Note that in here, simulations are always performed inside rectangular physical domains. This choice allows particular optimisations in comparison with

[1]OMNI web service, NASA: https://omniweb.gsfc.nasa.gov/.

the algorithms required for unstructured meshes, such as the implementation of a single numerical scheme routine, independent of the refinement.

During the discretisation process, the choices of proper refinements are challenging. An overly coarse mesh may cause the solution to be not adequately represented, especially in cases that contain localised structures or steep gradients, causing loss of information. On the other hand, an exaggerated refinement leads to a considerable amount of unnecessary computations, wasting a lot of computational time and memory. In this context, the use of adaptive techniques proposes to overcome these limitations. These techniques maximise the efficiency of the simulation by using an adaptive mesh, which is more refined in the regions where the localised structures are present and is coarser in the smooth solution regions.

2.1 Adaptive Meshes

The MHD solver developed for this work uses an SAMR method to construct the adaptive meshes. The work of Berger and Oliger [3, 5] was the first to introduce this method. Subsequently, Berger and Colella [3, 4] proposed a simpler version, in which every mesh of the hierarchy must be aligned with cells of the next coarser level, allowing simpler interpolation operations. Bell, Berger and Saltzman [2] demonstrated that this version can be more efficient, especially with vector and super-scalar computers.

The strategy of the SAMR methods to construct adaptive meshes is based on refinement criteria that measure the local smoothness of the solution in every mesh element. If the result of the criteria exceeds a predetermined value ϵ, the mesh element is flagged for refinement. After that, the flagged elements are overlaid by submeshes, i.e. patches, of finer cells that are refined by a factor of r. Note that the presently implemented multiresolution criterion only uses a refinement factor of $r = 2$; however, AMROC principally permits arbitrary integer values for other refinement criteria.

The process of overlaying more refined submeshes over coarser meshes produces a hierarchy of embedded level domains, i.e. the domain covered by a finer refinement level is also covered by the next coarser level. This hierarchy can be expressed as the sequence of meshes $\mathcal{M}^\ell$, where ℓ is associated with the refinement levels $\ell = 0, 1, \ldots, \mathrm{L}$. Each mesh $\mathcal{M}^\ell$ has its spatial mesh widths denoted as Δx^ℓ, Δy^ℓ and Δz^ℓ so that they present a constant ratio r between adjacent levels. Considering that the meshes in this hierarchy are embedded, the physical domain covered by each mesh $\mathcal{M}^\ell$ follows the property $\mathcal{M}^{\mathrm{L}} \subset \mathcal{M}^{\mathrm{L}-1} \subset \cdots \subset \mathcal{M}^0 = \Omega$, where the base mesh $\mathcal{M}^0$ covers the entire physical domain Ω. Each one of these meshes is divided into a set of non-overlapping rectangular submeshes $\mathcal{M}^\ell_m$ so that $\mathcal{M}^\ell := \cup_{m=1}^{M_\ell} \mathcal{M}^\ell_m$ with $\mathcal{M}^\ell_{m_1} \cap \mathcal{M}^\ell_{m_2} = \emptyset$, $m_1 \neq m_2$, where M_ℓ is the number of submeshes, or patches, used to represent the mesh $\mathcal{M}^\ell$.

One of the advantages of this mesh hierarchy is allowing a single implementation of the numerical scheme to be executed on each patch, independent of the refinement level. Thereby, the time evolution process can be performed for every patch individually. However, to compute fluxes along the boundaries of every patch requires a solution of the cells in adjacent patches. This restriction compromises the patch independence to perform the time evolution. In order to overcome this limitation, the patch structure is complemented with extra auxiliary cells, called ghost cells, at their boundaries, allowing the boundary values to be stored in the same data structure as the submesh.

The values in the ghost cells must be set or updated before and during the time evolution. For that, the ghost cells are divided into three cases regarding their position. At the physical boundary, the ghost cells are updated according to the problem boundary conditions. If there exists an adjacent patch on the same level, the ghost cells are updated by copying the solution from the neighbour. If the adjacent patch is, however, coarser, the ghost cells are interpolated from the solution in the coarser level using a multi-linear interpolation. In order to ensure discrete conservation in the fluxes between patches of different levels, the SAMR approach eventually replaces the coarse-cell fluxes by averaged fine-mesh fluxes, hence modifying the numerical stencil on the coarse mesh, cf. [14, 15] for AMROC-specific implementation details.

Wavelet Coefficients as Refinement Indicator

Wavelet theory shows that the decay of the wavelet coefficients estimates the local regularity of the solution [11]. Therefore, we can use such coefficients to predict where we could not have an adequate local approximation, and consequently, we need to improve the refinement. This can be used to determine dyadic refinement (i.e. $r = 2$) as these coefficients indicate regions of steep gradients or discontinuities [20, 37]. In this context, our proposed MHD solver implements ideas based on the adaptive multiresolution method introduced by Harten [27] to use as the refinement criteria. For that, the solver uses the multiresolution operations to predict the expected solution at a finer scale based on the solution from a coarser scale. Then, this prediction is compared with the actual solution. Besides this criterion, the MHD solver also allows using the criteria discussed in [15, 19].

Our underlying idea of applying multiresolution techniques for mesh adaptation is based on representing the numerical solution in two or more different resolution levels. Compression of the number of mesh cells, corresponding to coarsening the mesh locally, can be obtained by checking what happens between subsequent mesh resolutions [18]. For instance, for a discrete solution of a finite volumes discretisation, we consider as initial cell average data $\overline{\mathbf{Q}}^{\ell+1}$ at level $\ell + 1$. The transformation of these data $\overline{\mathbf{Q}}^{\ell+1}$ into an equivalent multiscale representation in one decomposition level is

$$\overline{\mathbf{Q}}^{\ell+1} \underset{\text{prediction}}{\overset{\text{projection}}{\rightleftharpoons}} \overline{\mathbf{Q}}^{\ell+1}_{MR} = \{\overline{\mathbf{Q}}^{\ell}\} \cup \{\mathbf{d}^{\ell+1}\},$$

where the set $\mathbf{d}^{\ell+1}$ contains the information between the two consecutive levels ℓ and $\ell + 1$, and $\overline{\mathbf{Q}}^{\ell}$ stores a smoothed version of the original numerical solution $\overline{\mathbf{Q}}^{\ell+1}$. The data at the highest resolution level are transformed into a set of coarser scale approximations plus a series of prediction errors corresponding to wavelet coefficients $\mathbf{d}^{\ell}$ in a multiresolution analysis. In order to perform the MR method with finite volume data, operations for projection and prediction are required, where the cell values are local averages.

Patch Creation
In SAMR methods, once the refinement criteria flag the coarser cells that require refinement, these cells are clustered into blocks by using a dedicated algorithm. Then, these blocks are used to construct patches of a finer refinement level. The AMROC framework uses the clustering algorithm proposed in [2], which was inspired by image detection techniques. This clustering algorithm is illustrated in Fig. 1. Let Υ_i be the number of flagged cells, i.e. *signatures*, in the ith row or column of cells on the current mesh $\mathcal{M}^{\ell}$. As the first step, this method splits the domain in every row and column where $\Upsilon_i = 0$. In the second step, the cuts are placed where the discrete second derivative $\Delta = \Upsilon_{i+1} - 2\,\Upsilon_i + \Upsilon_{i-1}$ crosses zero, starting from the steepest zero crossing and then using the lowest ones recursively. This step is repeated until the ratio among all cells and flagged ones in every new submesh is above the prescribed clustering ratio ϑ. In principle, $\vartheta \in [0.00, 1.00]$; however, in practice, typically $\vartheta \in [0.80, 0.99]$.

Time Evolution Strategy
As the adaptive mesh is defined, the time evolution for simulations with adaptive meshes presents minor challenges regarding stability and conservation. Obviously, the time step parameter Δt needs to satisfy the Courant–Friedrichs–Lewy (CFL) condition in every patch. In this context, the time evolution process using adaptive meshes may be performed with two different approaches [15]. One is based on a global time stepping where the patches of every level are updated with a Δt that satisfies the CFL condition on the finest meshes. The other strategy is a refinement-based recursive time stepping where Δt_{ℓ} varies between levels in the same proportion as their spatial refinement. This latter strategy requires the solution in patches of different refinement levels to be available at different time instants, which is handled in the SAMR method by constructing time-interpolated coarse level data as a boundary condition for interior fine level patches.

2.2 *Implementation Aspects*

AMROC is a freely available framework[2] that uses object-oriented programming concepts in the C++ language to support the numerical simulation of partial

[2]AMROC webpage: http://www.vtf.website/asc/wiki/bin/view/Amroc.

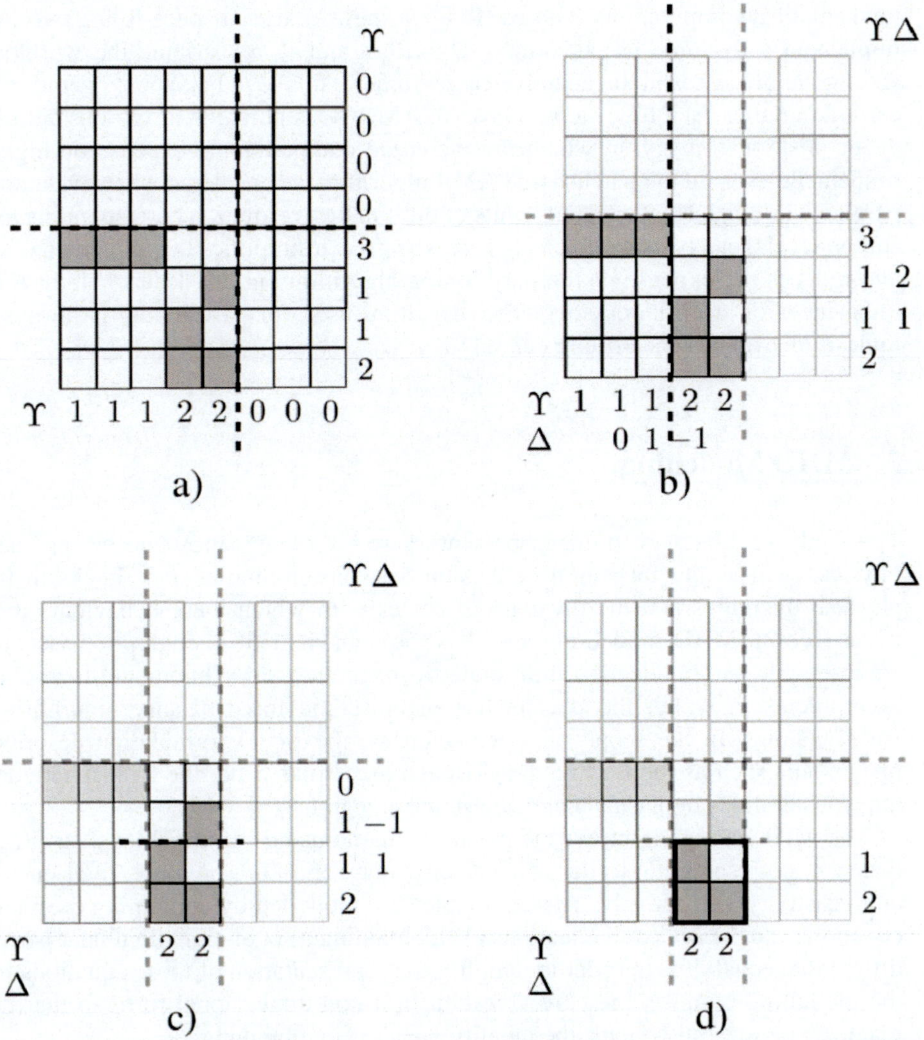

Fig. 1 Cell clustering algorithm. The value Υ_i associated with each row and column is defined as the number of flagged cells in that row or column, and $\Delta_i = \Upsilon_{i+1} - 2\Upsilon_i + \Upsilon_{i-1}$. (**a**) 1st step: mesh division where $\Upsilon = 0$. (**b**) 2nd step: division of the white mesh where the value Δ_n presents the highest variation between adjacent values with opposite signals. (**c**) Recursive execution of the 2nd step over the new subdivisions. (**d**) Stop criteria: percentage of flagged cells in the subdivision surpass the predefined clustering ratio ϑ

differential equations using adaptive methods. In this framework, presented in [15], the SAMR core contains about 46,000 lines of code C++ and around 6000 lines of code for visualisation and data conversion routines. Besides being written in the C++ language, the framework uses the FORTRAN language to perform mesh operations, such as prolongation and restriction, due to its better performance in mathematical computations. The AMROC framework uses a space-filling curve to implement a dynamic re-partitioning algorithm and to redistribute the workload among the processes in the adaptive cases. In here, this load balancing operation is carried out after each time step on level 0. AMROC is pursuing a rigorous domain decomposition strategy, in which the increased computational expense on higher refinement levels in the patch-based AMR algorithm is considered when evaluating parallel workload. However, only units of the smallest resolution corresponding to a cell on level 0 can be considered [14]. This approach simplifies the implementation and reduces the expense of the partitioning algorithm but can lead to slight load imbalances on deep hierarchies. The algorithm used for partitioning is always a multi-dimensional space-filling curve [13, 15].

3 MHD Modelling

The study of MHD phenomena presents a series of demands concerning their physics, such as the formation of instabilities, discontinuities and shocks in the physical quantities and the diversity of scales with which these behaviours may occur [6, 32]. MHD modelling describes the behaviour of a single, non-viscous, compressible and conducting fluid under a magnetic field. This model is applied in problems in which the plasma has macroscopic force balance, equilibrium and dynamics. In the scope of space sciences, the MHD formulation describes phenomena such as the Earth magnetosphere, the solar wind, the heliosphere and many instabilities in plasma that occur in those regions.

The MHD model describes the plasma dynamo using the variables ρ, $\mathbf{u}$, $\mathbf{B}$, p, $\mathcal{E}$ and η, corresponding to density, velocity, magnetic flux, pressure, total energy and resistivity, respectively. These variables are modelled by combining the Euler equations and the Maxwell equations [32], obtaining a set of eight nonlinear partial differential equations. In order to simplify the representation of these equations for the modelling purpose, they are rewritten in a non-dimensional form so that the magnetic permeability yields the identity $\mu = 1$ [25], obtaining

$$\frac{\partial \rho}{\partial t} + \nabla \cdot (\rho \mathbf{u}) = 0, \tag{2a}$$

$$\frac{\partial (\rho \mathbf{u})}{\partial t} + \nabla \cdot \left[\rho \mathbf{u}\mathbf{u} + \left(p + \frac{\mathbf{B} \cdot \mathbf{B}}{2}\right)\mathbb{I} - \mathbf{B}\mathbf{B}\right] = \mathbf{0}, \tag{2b}$$

$$\frac{\partial \mathcal{E}}{\partial t} + \nabla \cdot \left[\left(\mathcal{E} + p + \frac{\mathbf{B} \cdot \mathbf{B}}{2}\right)\mathbf{u} - (\mathbf{u} \cdot \mathbf{B})\,\mathbf{B} + [\eta\,(\nabla \times \mathbf{B})] \times \mathbf{B}\right] = 0, \tag{2c}$$

$$\frac{\partial \mathbf{B}}{\partial t} + \nabla \cdot \left[\mathbf{u}\mathbf{B} - \mathbf{B}\mathbf{u} + \eta\left((\nabla \mathbf{B})^T - \nabla \mathbf{B}\right)\right] = 0. \tag{2d}$$

This system is completed by the definition of total energy, which is the combination of the hydrodynamic and magnetic energies, i.e.

$$\mathcal{E} = \frac{p}{\gamma - 1} + \rho\frac{\mathbf{u}\cdot\mathbf{u}}{2} + \frac{\mathbf{B}\cdot\mathbf{B}}{2}, \tag{3}$$

where γ is the adiabatic index. Furthermore, the plasma must satisfy the Gauss's law for the magnetism $\nabla\cdot\mathbf{B} = 0$, i.e. the divergence over the magnetic field is constant and zero over time, not allowing the formation of magnetic monopoles (since it is a nonphysical behaviour).

Numerical Aspects

The numerical simulation of the MHD equations presents several inherent challenges, for instance, the operator $\nabla\cdot(\nabla\times\cdot)$ over a vector quantity is zero, which implies, in our study, that $\partial_t\,(\nabla\cdot\mathbf{B}) = 0$. So, $\nabla\cdot\mathbf{B}$ is a constant value, but as earlier mentioned, magnetic monopoles can not be formed. However, in many numerical methods, due to numerical approximation errors, the term containing $\nabla\cdot(\nabla\times\cdot)$ does not indeed result in zero. Hence, $\nabla\cdot\mathbf{B}\neq 0$, so that the MHD system does not satisfy Gauss's law as expected. The loss of this constraint causes spurious behaviours in the numerical solution, creating magnetic monopoles. This phenomenon leads to numerical instabilities as discussed, for instance, in [7, 46], and more recently in [29].

To deal with this problem, the AMROC-MHD module implements the Generalised Lagrangian Multiplier (GLM) formulation for the MHD equations [12]. In general, GLM methods are used to maximise or minimise a function under some constraints. In this context, the GLM method is used to maximise the induction equation while imposing $\nabla\cdot\mathbf{B} = 0$. This is done by coupling a differential operator $\mathcal{D}$ to the Gauss's law so that $\mathcal{D}(\psi) + \nabla\cdot\mathbf{B} = 0$. Then, the solution ψ is coupled to the induction equation, obtaining

$$\frac{\partial\mathbf{B}}{\partial t} + \nabla\cdot\left[\mathbf{u}\mathbf{B}^{\mathbf{T}} - \mathbf{B}\mathbf{u}^{\mathbf{T}} + \eta\left((\nabla\mathbf{B})^{T} - \nabla\mathbf{B}\right) + \psi\mathbf{I}\right] = \mathbf{0}. \tag{4}$$

In particular, the MHD simulations in this work use the parabolic–hyperbolic divergence cleaning approach as proposed in [12] and updated in [34]. This correction is characterised by the operator $\mathcal{D}(\psi) = \frac{1}{c_p^2}\psi + \frac{1}{c_h^2}\frac{\partial\psi}{\partial t}$, with c_p and $c_h\in(0,\infty)$, where the parameter $c_h = \max\left[\sigma\frac{\Delta h}{\Delta t},\, \max\left(|u_i|\pm c_f\right)\right]$, with Δh being the minimal value of the mesh sizes in each direction, σ the Courant number, u_i the velocity of the ith component, c_f the fast magneto-acoustic wave velocity of the MHD model, and the c_p value is defined in terms of the parameter $\alpha_p = \frac{\Delta h\, c_h}{c_p^2}$ for $\alpha_p\in[0,1]$. Applying this operator into the modified Gauss's law, the equation that describes the evolution of ψ is obtained as

$$\frac{\partial\psi}{\partial t} + c_h^2\nabla\cdot\mathbf{B} = -\frac{c_h^2}{c_p^2}\psi. \tag{5}$$

The choice of this operator results in a method that transports and diffuses the components of $\nabla \cdot \mathbf{B}$ to the boundaries [12]. Besides not eliminating these components completely, this method is capable of maintaining the accuracy and the stability of the solution at a very low computational cost.

Moreover, the AMROC-MHD module also implements the Extended GLM-MHD formulation (EGLM-MHD), as described in [12], that includes Powell's source terms into the GLM-MHD model. However, we will not discuss EGLM experiments herein as the GLM approach presents solutions that are mathematically more rigorous in the sense of conservation laws.

The discussed GLM-MHD system is completed by suitable initial and boundary conditions as presented in the numerical experiments section. Note that the variable ψ is always initialised as zero for the entire domain.

The AMROC-MHD solver implements the HLL [28] and HLLD [35] fluxes using a two-stage second-order accurate Runge–Kutta method to perform the flux computations of the MHD equations. Furthermore, the AMROC-MHD module uses MUSCL (Monotone Upstream-Centered Scheme for Conservation Laws) [50] as a high-resolution scheme. This approach is based on the usage of a slope limiter that consists of piece-wise linear reconstructions to extrapolate each variable $\mathbf{q}$ of the solution to the left and right boundaries of the cell. These extrapolations are performed as presented in [45].

Besides the GLM formulation, in order to ensure that the solution does not develop spurious oscillations around discontinuities or shocks, the AMROC-MHD module implements the following slope limiter functions to be used in a MUSCL scheme: Minmod [41], Monotonized Central (MC) [49], Superbee [41], van Albada [51], van Leer [48] and Koren [31].

4 Experiments and Discussions

This section presents the results of the adaptive simulations of two benchmark cases for a three-dimensional ideal MHD and a two-dimensional resistive MHD model, respectively. These benchmarks represent significant phenomena that appear in the physics of the space environment involving the Earth magnetosphere model. The experiments aim to quantify the scheme accuracy in the $\mathbb{L}_1$-norm, the memory compression and the CPU-time gain concerning a non-adaptive mesh with the same refinement as the finest level, as discussed in [16, 18]. Finally, a more complex configuration setup involving the Earth magnetosphere is discussed.

All experiments are conducted using a Cartesian mesh, the HLLD numerical flux introduced in [35], and the MinMod limiter as discussed, for instance, in [45]. The computations were run in parallel using nodes of a recent GNU/LINUX computer cluster that provides 20 cores with shared memory per node.

4.1 Magnetic Shock-Cloud (MSC)

The magnetic shock-cloud problem is a benchmark test that verifies the performance of the numerical scheme when dealing with super-fast flows [46]. It describes the disruption of a high-density magnetic cloud by a strong shock wave. For that, an advancing plasma is considered, which causes a shock with a stationary state that contains a high-density cloud. This is a conceptual problem, not yet computationally exploited in the space science literature. These simulations are performed inside the computational domain $[0, 1]^3$ with outlet boundaries, using the GLM factor $\alpha_p = 0.4$ and $\eta = 0$ (ideal MHD case). The time steps are performed under the Courant number $\sigma = 0.4$ until the final time $t = 0.06$. The initial states of both the advancing (delimited by $x < 0.05$) and steady ($x > 0.05$) plasma regions are described in Table 1 for the adiabatic constant $\gamma = \frac{5}{3}$, with non-dimensional quantities compatible with the defined MHD model. The density solution in the steady plasma ρ^0 is given by value $\rho^0 = 10$ if the coordinates are inside the cloud with centre in $(0.25, 0.5, 0.5)$ and radius $r = 0.15$. Otherwise, this density is set as $\rho^0 = 1$. Figure 2 shows a slice representation of the density initial condition configuration and the solutions at the instants $t = 0.03$ and $t = 0.06$ alongside with the respective adaptive meshes and the mesh distribution among the 48 processors used for these simulations.

The figure containing the refinement of the adaptive mesh is interpreted such that the blue regions of the domain represent the coarsest scale, while the red regions represent the most refined scale. The adaptive mesh localises structures in the solution, such as the bow shock, the centre of the explosion area and the tail. The visualisation of the mesh distribution per processor is a complicated task in this three-dimensional case. However, we can roughly estimate the form of this distribution using planar cuts, cf. right graphic of Fig. 2, so that each colour represents the sub-domain evolved by each processor. The workload balance algorithm is expected to create small sub-domains in the most refined regions, while the coarser regions are evolved using less processors. This can be observed in the initial condition, where the refinement concentrates in the shock wave and in the sphere borders, and hence, more processors are being used in these regions, in contrast with the region with $x > 0$, which is predominantly being evolved by the same processor. In the instant $t = 0.03$, the processors maintained their concentration around the cloud region and the yz-plane, where the shock wave is located, while the region $x > 0$ is still evolved by few processors. At the final instant $t = 0.06$, the processors are more spread into the domain, but a concentration is still visible in the cloud and tail regions. These graphics indicate an excellent distribution in the regions where more refinement is desired.

Table 1 MSC: initial conditions. This IC is complemented by the values $u_y = u_z = B_x = 0$

	ρ	p	u_x	B_y	B_z
$x < 0.05$	3.86859	167.34500	11.25360	2.1826182	−2.1826182
$x > 0.05$	ρ^0	1	0	0.56418958	0.56418958

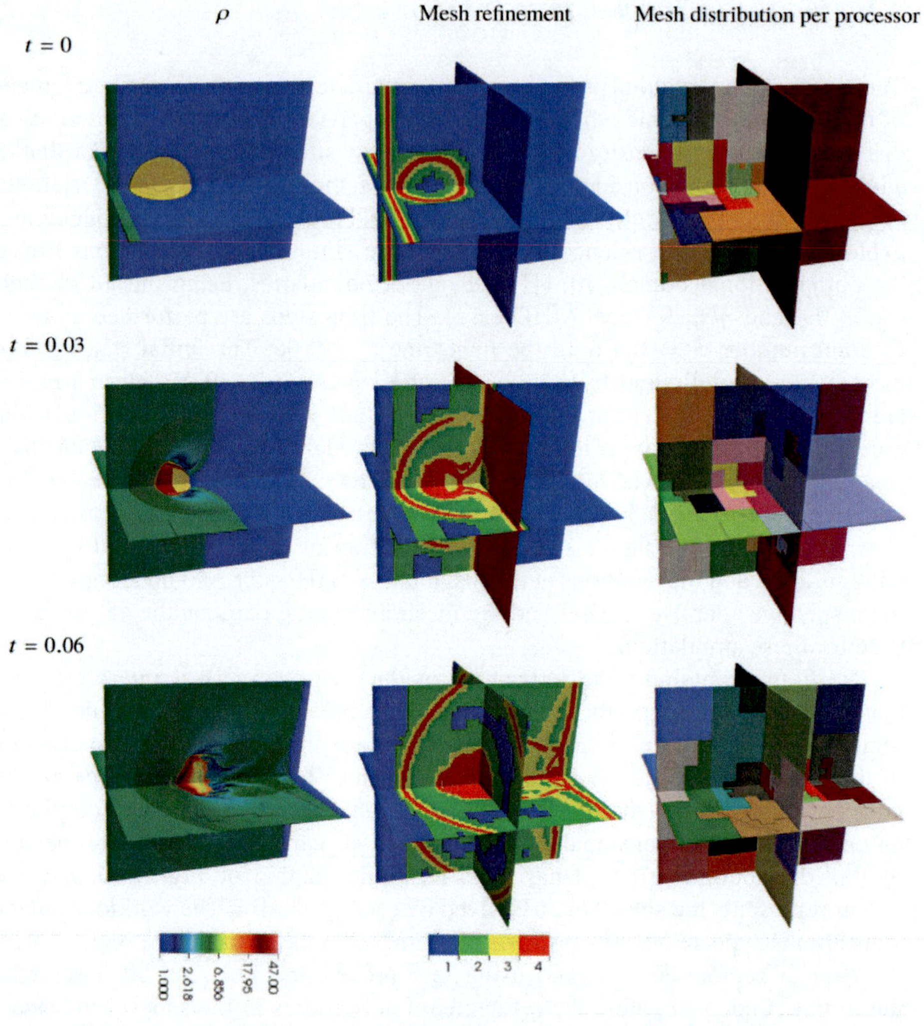

Fig. 2 MSC: Results for density ρ, the adaptive mesh refinement and distribution per processors. This simulation is performed using $\mathrm{L} = 4$ refinement levels with a base mesh size of 128^3 cells, MinMod limiter, threshold value $\epsilon = 0.025$ and 48 processors until $t_{\text{end}} = 0.06$

Table 2 presents the results of the AMR simulations using a refinement threshold value $\epsilon = 0.025$ with a number L of extra refinement levels, so that the finest level allowed corresponds to a uniform mesh size of 512^3 cells. This table also presents the $\mathbb{L}^1$-norm error for pressure, the CPU time, the number of cells used in the discretisation at the final instant and the number of patches used in the grid hierarchy for each case. For these parameters, the simulation with $\mathrm{L} = 3$ refinement levels presented a better structure localisation than the one with $\mathrm{L} = 2$, due to the number of cells and patches required. This resulted in a significative reduction in the CPU time (around 13%) with a small increase in the overall $\mathbb{L}_1$-error in the order of 10^{-4}. Moreover, the simulations with $\mathrm{L} = 3$ and 4 refinement levels presented

Table 2 MSC: errors in pressure p, memory consumption (number of cells and patches used) and CPU time obtained by using several refinement levels L using MinMod limiter at $t_{\text{end}} = 0.06$ with a threshold value $\epsilon = 0.025$

Mesh			Accuracy (p)	Cells		Patches	CPU time	
Size	Base	L	$\mathbb{L}_1$-error ($\times 10^{-3}$)	#	%	#	Min	%
512^3	64^3	4	2.481	34,032,848	25	10,812	442.8	32
512^3	128^3	3	2.4176	33,967,120	25	11,548	458.9	33
512^3	256^3	2	2.1898	40,190,312	30	22,420	636.4	46

Table 3 MSC on uniform and MR adaptive mesh computations using MinMod limiter

	Uniform mesh		Adaptive mesh, $\epsilon = 0.025$, $\vartheta = 0.80$			
Mesh (size)	# Cells	CPU time (s)	Base mesh (size)	# Patches	# Cells	CPU time (s)
256^3	16,777,216	4563.6	32^3	148	5,917,288	1468.5
512^3	134,217,728	74,185.5	32^3	894	27,444,592	11,350.3
1024^3	1,073,741,824	–	32^3	847	28,714,656	117,731.4

similar number of cells and patches in their adaptive meshes. Consequently, these cases presented similar CPU time. This indicates that the fourth refinement level has barely been used, causing the number of cells and patches to be similar. Nevertheless, the small reduction in the CPU time by 1% obtained by the L $=$ 4 simulation, in relation to the L $=$ 3 simulation, increased the overall $\mathbb{L}_1$-error by the order of 10^{-5}. Therefore, the result with L $=$ 4 is considered to present the best gain considering precision and CPU time.

In Table 3, we present a comparison between the uniform and an adaptive computation with a refinement threshold value $\epsilon = 0.025$ and a clustering ratio $\vartheta = 0.80$ for meshes with different finest refinements and using the same base mesh 32^3. The gain of the adaptive computations sharply increases with the enlargement of the mesh. In particular, for the adaptive computation, we use less than 3% of the number of cells for computation related to a uniform mesh 1024^3 and spend a CPU time close to the computation of the uniform mesh 512^3. Moreover, in the highest resolved computation, the uniform mesh exceeds our cluster memory, and we therefore cannot provide the CPU time for this computation. The numbers of patches are almost similar among the two refined meshes; however, the most refined mesh presents a slightly smaller number of patches, which seems to indicate a better localisation of the structures.

Considering the finest adaptive mesh 1024^3, Table 4 shows the comparison with different cluster parameters and base meshes taking into account the number of cells and the CPU time. In both cases, the clustering ratio $\vartheta = 0.99$ presented the largest number of patches, which is expected. In contrast to what we observed in the two-dimensional case [19] that these results produced less computational effort and number of cells. This could be related to the localisation of the structures in the solution in this three-dimensional case. Therefore, in this case, a clustering ratio $\vartheta = 0.99$ presented better results.

Table 4 MSC: cell-cluster comparison for MR adaptive computations considering different clustering ratios ϑ and base meshes, with $\epsilon = 0.025$ and MinMod limiter

Mesh					CPU
Size	Base	ϑ	# Patches	# Cells	time (s)
1024^3	16^3	0.99	392	28,730,312	158,152.4
1024^3	16^3	0.80	351	28,991,200	180,701.6
1024^3	32^3	0.99	868	28,476,728	116,871.8
1024^3	32^3	0.80	847	28,714,656	117,731.4

4.2 Magnetic Reconnection (REC)

As a relevant question of space science, the magnetic reconnection problem is described in [30] as the merging of the magnetic field lines from two predominantly opposing magnetic fields, liberating a considerable amount of energy and redirecting the direction of particle flows. This type of phenomenon is common in solar physics and is highly studied due to the effects of the interaction between the Earth's magnetic field and the interplanetary magnetic field that creates complicated space environment processes. This test aims to verify the implementation of the resistive terms in MHD, once the resulting effects can be connected with the process responsible for the morphological transition of the magnetic field lines and changing in the plasma's flux, as presented in [42].

The problem is initialised considering two different states, divided at $x = 0$, with a small transition gap between them. The two states have magnetic fields that present opposing orientation over direction, and the reconnection occurs inside a small region inside the transition gap where there is a small resistivity. The computational domain for this problem is $\left[-\frac{1}{2}, \frac{1}{2}\right] \times [-2, 2]$ with Dirichlet boundary conditions. Inside this domain, the resistivity is defined as

$$\eta(x, y) = \frac{\eta_0}{4}\left[1 + \cos(10\pi x)\right]\left[1 + \cos(2.5\pi y)\right]$$

if (x, y) is inside the sub-domain $[-\mathrm{L}_r, \mathrm{L}_r] \times \left[-\frac{1}{5}, \frac{1}{5}\right]$, and is zero elsewhere with the parameters $\eta_0 = 6 \cdot 10^{-4}$ and $\mathrm{L}_r = 0.05$. The initial conditions of the physical quantities are $\rho = 1$, $p = 0.1$ and $\mathbf{u} = \mathbf{0}$. The components of the initial magnetic field are given according to its corresponding state, as presented in Table 5, with non-dimensional quantities compatible with the defined MHD model. The simulations of this problem are performed using the adiabatic constant $\gamma = \frac{5}{3}$ and the HLLD Riemann solver combined with the MC limiter. The parabolic–hyperbolic correction uses the factor $\alpha_p = 0.4$. All simulations are performed under Courant number $\sigma = 0.4$ until the final time $t = 2.5$.

Table 6 presents the error in pressure p using the $\mathbb{L}_1$-norm, the CPU time and the number of cells and patches used in the adaptive mesh for simulations with several refinement levels. These simulations are performed using the threshold value of $\epsilon = 0.001$ for the MR refinement criteria. The number of levels used

Table 5 REC: magnetic field initial condition

Region	B_x	B_y	B_z
$x < -\mathrm{L}_r$	0	−1	0
$x > \mathrm{L}_r$	0	1	0
Transition zone	0	$\sin\left(\frac{\pi}{2\mathrm{L}_r}x\right)$	$\cos\left(\frac{\pi}{2\mathrm{L}_r}x\right)$

Table 6 REC: errors in pressure p, memory consumption and CPU time with refinement levels L

Mesh			Accuracy (p)	Cells		Patches	CPU time	
Size	Base	L	$\mathbb{L}_1$-error	#	%	#	Min	%
1024 × 2048	128 × 256	4	0.0231	928,932	44	852	27.7	29
1024 × 2048	256 × 512	3	0.0226	895,292	42	809	48.1	51
1024 × 2048	512 × 1024	2	0.0050	856,968	40	466	48.3	51

in each simulation is configured such that the most refined scale corresponds to a 1024 × 2048 mesh. The simulation that presented the best results is obtained with L = 2, which corresponds to a reduction of 49% of the CPU time, while maintaining an error in the order of $5 \cdot 10^{-3}$. In that case, the gain is roughly four times concerning the simulation with L = 4. Besides, this case presented the lowest number of cells concerning the uniform mesh simulation and also the lowest number of patches. Furthermore, these adaptive cases required only around 40% of the cells of the uniform mesh simulation.

Figure 3 presents the solution for p, and the adaptive mesh for the simulation with 4 levels. Showing the adaptive mesh refinement, the figure can be interpreted such that the brightest regions of the domain represent the coarsest scales, while the darkest regions correspond to the most refined scales. Physically, this figure represents a snapshot of the magnetic reconnection process. One can identify the interface between the domains of opposing orientation connected to the field line merging processes. Furthermore, there is a convergence of plasma in one direction and divergence in a perpendicular direction, as can be seen in the velocity plot. These adaptive results are in agreement with the MHD solutions presented in [30], and the modelling results are in agreement with the underlying physics, as considered in [42].

Table 7 presents a breakdown of the most computationally costly tasks of the adaptive REC simulations with different number of refinement levels. In comparison with the uniform mesh simulation, the *Integration* costs of the adaptive simulations exhibit a significant reduction, as expected from the lower number of integrated cells. Moreover, the *Boundary* costs also show a reduction for three and four refinement levels, while costs for not explicitly timed operations (*Misc*) reduce as more levels are included. The *Output* production cost is generally insignificant, and however, it also decreases for larger level number. Besides these cost reductions,

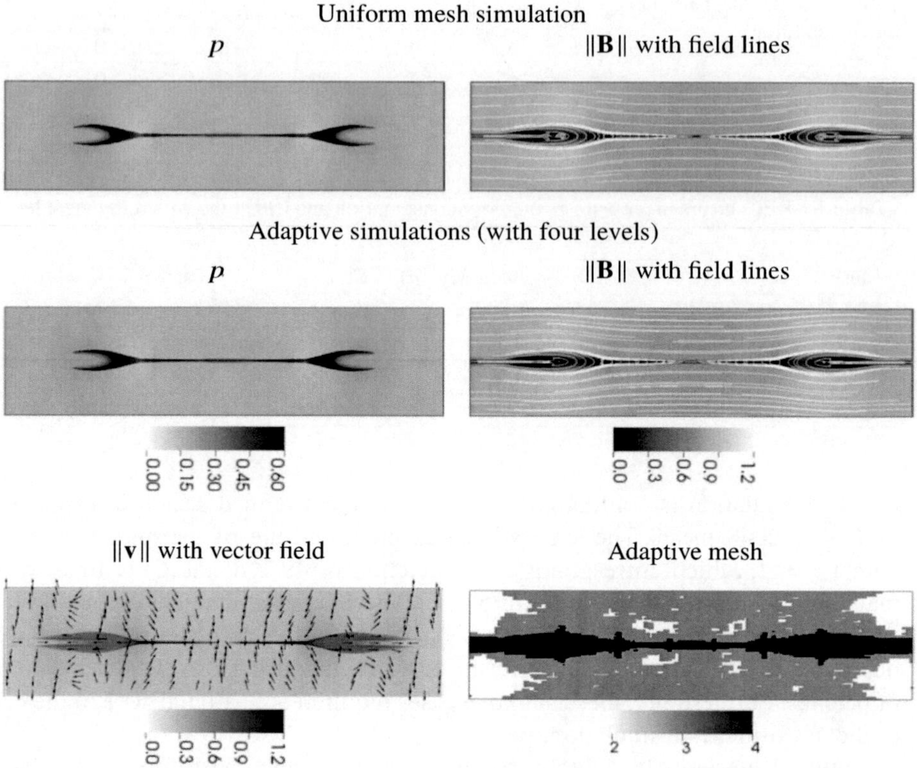

Fig. 3 REC solution for uniform mesh (p, and $\|\mathbf{B}\|$), and the respective adaptive simulations for (p, $\|\mathbf{B}\|$, and $\|\mathbf{v}\|$) with their adaptive meshes. This simulation is performed using four refinement levels with threshold value $\epsilon = 0.001$ and 24 processors. These figures are presented using the y-axis in the horizontal direction

Table 7 REC: breakdown of the CPU time, in seconds, spent in main computation tasks for different numbers of refinement levels L using $\epsilon = 0.001$

	Refinement levels			
Main task	One	Two	Three	Four
Integration	3756.8	931.5	747.2	737.9
Boundary	796.1	852.9	591.0	503.5
Memory restart	142.0	29.5	11.0	5.7
Recomposition	–	500.7	1141.0	228.2
Remeshing	–	24.1	21.4	21.4
Misc	518.6	359.8	203.0	116.6
Output	3.8	6.4	5.4	1.2

the *Recomposition* and *Remeshing* tasks associated with the adaptive simulations present a significant cost, especially for the three-level simulation. Nevertheless, even considering these costs, the adaptive simulations require less CPU time than the uniform mesh simulations.

4.3 Magnetosphere (MAG)

Initially as conceived from the incidence of electrically charged solar particles, the Chapman–Ferraro model proposed that the close Earth's space environment was an empty region that avoids the presence of those particles [8, 9]. Nowadays, governed primarily by the geomagnetic field, a much more complicated electrodynamics region develops surrounding the planet, designated as magnetosphere [42], populated by plasmas.

As application cases for the solver, the first analysis deals with a predominantly northward-oriented interplanetary magnetic field (IMF) causing—in principle—a geomagnetically closed frontal magnetosphere, while the second case takes into account the southward-oriented IMF, which causes a geomagnetically open frontal magnetosphere. The main modelling ideas are described in details in [39, 40] and references therein. For these studies, the close Earth's environment is considered basically as a sphere with density and pressure constant in time and containing a magnetic dipole. This region is connected to an outer region where the geomagnetic field is compressed or stretched by the solar wind, defining a region where the well-known phenomenon of magnetic reconnection can occur.

The physical model of this problem requires changes in the resistive MHD formulation presented in Eqs. 2. These changes consist of the inclusion of an external gravity field, an artificial viscosity over the density and pressure field to reduce the MHD fluctuation that arises from the unbalanced forces in the initial condition, and a modification in the Ampère's law so that $\mathbf{J} = \nabla \times (\mathbf{B} - \mathbf{B_d})$, where $\mathbf{B_d}$ is the intrinsic dipole magnetic field of the Earth [42, p. 223]. The reason to subtract the dipole field from the Ampère's law is supported to the expectation of a significant electric current to be generated in the frontal interface layer between the two media, i.e. the interplanetary space and the outer terrestrial region [38–40]. These modifications imply the inclusion of the source terms $D\nabla^2\rho$ in the continuity equation 2a, $\rho\mathbf{g} + \Phi + \mathbf{u}D\nabla^2\rho + \mathbf{B} \times (\nabla \times \mathbf{B_d})$ in the momentum equation 2b, and $\rho\mathbf{u}\cdot\mathbf{g} + \frac{D_p\nabla^2 p}{\gamma-1} + \frac{\|\mathbf{u}\|^2}{2}D\nabla^2\rho + \mathbf{u}\cdot\Phi + \eta\|\nabla\times\mathbf{B_d}\|^2 + (\nabla\times\mathbf{B_d})\cdot(\mathbf{u}\times\mathbf{B} - \eta\nabla\times\mathbf{B})$ in the total energy equation 2c. The diffusive terms in this model are given in [39] as $D = D_p = \frac{\mu}{\rho_{SW}} = 0.02$ and $\Phi = \mu\nabla^2\mathbf{u}$, where $\rho_{SW} = 5.00 \cdot 10^{-4}$ corresponds to the typical value (5 n/cc) for the solar wind density. Moreover, we use the following induction equation for the magnetic field:

$$\frac{\partial \mathbf{B}}{\partial t} + \nabla \cdot \left[\mathbf{uB} - \mathbf{Bu} + \eta\left((\nabla\mathbf{B})^T - \nabla\mathbf{B} - (\nabla\mathbf{B}_d)^T + \nabla\mathbf{B_d}\right) + \psi\mathbf{I}\right] = \mathbf{0}. \quad (6)$$

Physical Quantities

The physical quantities included in the model are described by a function related to the distance to the centre of Earth, represented by ξ $(= \sqrt{x^2 + y^2})$ defined with the Earth located in the xy-plane origin. This two-dimensional formulation considers the x-axis in the Sun–Earth direction crossing the Earth Equator, while the y-axis is orthogonal to the x-axis so that it contains the Earth north and south poles. This reference axis orientation is used for computational convenience in the code. Nevertheless, to be clear, in the domains of the space scientists that use an own reference coordinate system representation, this axis is considered to be the z-axis. Besides the initial conditions, the distance ξ is used to determine the external fields used in these simulations, such as the external gravity $\mathbf{g}$, defined by the vector field $\mathbf{g}(x, y) = -\frac{g_0}{\xi^2}[x,\ y,\ 0]$, where $g_0 = 1.35 \times 10^{-6}$; the line dipole magnetic field $\mathbf{B_d}$, given by $\mathbf{B_d}(x, y) = \left[-2xy\xi^{-4},\ \xi^{-4}\left(x^2 - y^2\right),\ 0\right]$, and the resistivity function $\eta(x, y) = \eta_0 \frac{w}{w+1}$, where the weight value reads $w = 30(\max\left[(\xi/16)^2 - 1,\ 0\right])^2$, as defined in Ogino's implementation.

Those non-dimensional quantities are obtained from a physical system of units in which the distance, magnetic induction and time are established based on the Earth's radius (RE $= 6.37 \cdot 10^6$ m). We provide some fundamental physical information used in this context. The Earth magnetic field at the equator is $3.12 \cdot 10^{-5}$ T, and the Alfvén transit time, a time required by the Alfvén wave to go through the equivalent of the Earth radius, is taken as 0.937 s. Based on those quantities, the pressure unit corresponds to $7.75 \cdot 10^{-4}$ N/m^2, the velocity unit to $6.80 \cdot 10^6$ m/s, the acceleration unit to $7.26 \cdot 10^6$ m/s^2 and the current density unit to $3.90 \cdot 10^{-6}$ A/m^2.

Initial and Boundary Conditions

The initial configuration for this problem consists in an approximation of the ionosphere, which describes the plasma in the Earth's neighbourhood, based on the initial condition proposed in [38–40]. This ionosphere is constructed so that its pressure and density are proportional to ξ. The initial magnetic field in this work agrees with the one proposed in the works [38, 39] by including an imaginary dipole at the coordinate $x = 2x_m$ so that $\mathbf{B}^0(x, y) = \mathbf{B_d}(x, y) + \mathbf{B_d}(2x_m, y)$, where x_m is the equilibrium point on the x-axis where the solar wind kinetic energy counterbalances the dipole's magnetic energy, i.e. the coordinate at which $\rho_{SW} u_{SW}^2 = \mathbf{B_d}^2$ for $y = 0$. This imaginary dipole, known as the image method in electrodynamics [42, p. 225-227], aids to produce an initial magnetic field that has its field lines compressed, and not crossing the line $x = x_m$. Its initial structures are constrained to the domain of the Earth (where $x \geq x_m$). This magnetic configuration is done to accelerate the convergence into a steady-state magnetosphere so that the Earth magnetic field does not cross the position immediately beyond the magnetopause, formed around $x = x_m$. In the domain of the interplanetary medium (where $x < x_m$), the initial magnetic field is set to be equal to the typical interplanetary magnetic field. This problem is simulated inside the physical domain $[-150, 450] \times [-150, 150]$ RE, which is complemented with Neumann boundary conditions so that the derivatives of the physical quantities are zero at the boundaries $x_e = 450$, $y_s = -150$ and $y_e = 150$. The boundary at $x_s = -150$ is used to prescribe the solar wind parameters.

The entire ionosphere initial condition and the typical solar wind parameters, used as boundary conditions, are given in Table 8, where $p_{00} = g_0(\gamma - 1)/\gamma = 5.40 \cdot 10^{-7}$ for $\gamma = 2$, $p_0 = 3.56 \cdot 10^{-8}$ (corresponding to a temperature of $T_{SW} = 2 \cdot 10^5$ K), $u_{SW} = 4.41 \cdot 10^{-2}$ (300 Km/s) and $B_{SW} = \pm 1.5 \cdot 10^{-4}$ (5 nT), in which

Table 8 Initial conditions and typical solar wind parameters

	ρ	p	u_x	u_y	u_z	B_x	B_y	B_z
$\mathbf{q}^0(x > x_m, y)$	$\max\left(\frac{1}{\xi^2}, 10^{-4}\right)$	$\max\left(\frac{p_{00}}{\xi}, p_0\right)$	0	0	0	B_x^0	B_y^0	0
$\mathbf{q}^0(x \le x_m, y)$	$\max\left(\frac{1}{\xi^2}, 10^{-4}\right)$	$\max\left(\frac{p_{00}}{\xi}, p_0\right)$	0	0	0	0	B_{SW}	0
$\mathbf{q}_{SW}(x_s, y, t)$	ρ_{SW}	p_0	u_{SW}	0	0	0	B_{SW}	0

the positive (negative) signal of B_{SW} is associated with the northward (southward) orientation of the magnetic field of the solar wind.

Furthermore, the physical domain also presents an internal boundary corresponding to a near-Earth region, i.e. a region defined indeed by the ionosphere and the plasmasphere (details can be seen in [26, p. 208–222, and p. 164–173, respectively]). Considering the Earth positioned at the origin, this internal boundary removes the points with $\xi < 16$ from the computational domain. In order to dampen out all perturbations near the ionosphere, the near-Earth neighbourhood is smoothed in relation with the initial condition after every time step by the operation $\mathbf{q}^{n+1} = f\,\mathbf{q}_*^{n+1} + (1-f)\,\mathbf{q}^0$, where $\mathbf{q}_*^{n+1}$ is the solution obtained after the time evolution and $\mathbf{q}^0$ is the ionospheric initial condition. The weight value f is computed as $f = \dfrac{\bar{f}^2}{\bar{f}^2+1}$, where $\bar{f} = 100\left(\max\left[\left(\frac{\xi}{16}\right)^2 - 1,\ 0\right]\right)^2$. As proposed, this function guarantees a smooth transition of quantities in a thin layer immediately surrounding the ionosphere boundary, which presents constant values. This approach avoids the MHD solver run stopping due to numerical instabilities.

Configuration of the Magnetosphere

The initial condition presented earlier describes the initial state of the ionosphere, without contemplating an initial state for the magnetosphere. Thus, before introducing inputs composed of realistic data from the interplanetary environment, an approximate configuration for the magnetosphere should be realised. The initial ionospheric configuration is simulated using the typical solar wind parameters $\mathbf{q}_{SW}$, presented in Table 8, in order to obtain an initial state for a magnetosphere in equilibrium. Figure 4 presents the density and, in white, the magnetic field lines. The panels in the figure contain the initial condition and intermediary states of the magnetosphere configuration before a stationary state is obtained using southward IMF. These configurations are presented alongside with its corresponding adaptive grids. An IMF orientation choice is made to match the orientation of the magnetic field at the beginning of the dataset. This choice is due to the Earth magnetic field, under typical northward and southward IMF, converging to different states, as presented in Fig. 5. In the lower panel in the figure, there is a frontal reconnection, and in the upper panel, reconnection does not occur in this position. The errors, CPU time and number of cells and patches of the adaptive simulations using two and three refinement levels are presented in Table 9 for both northern- and southern-oriented solar wind. The simulations with southern-oriented solar wind presented

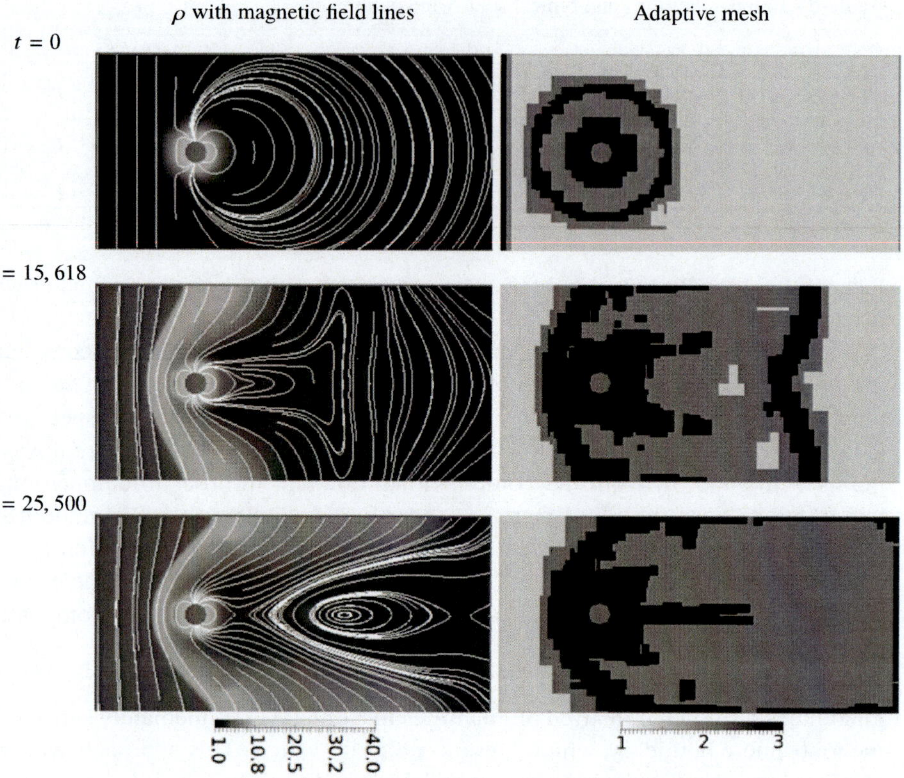

Fig. 4 Density solution (left) at various time instants during the southern solar wind steady-state magnetosphere construction with its correspondent adaptive mesh (right) with finest level corresponding to a 512×256 mesh and threshold value $\epsilon = 0.05$. Magnetic field lines represented in white evolving from a southward orientation

better performance due to its simpler mesh configuration, as illustrated in Fig. 5. Besides the dipole, which is being compressed on the solar wind side while being elongated downstream for both cases, these steady-state magnetospheres differ by potential locations of the magnetic reconnection process. These simulations with typical solar wind parameters are assumed to have converged to steady state at the instant $t = 172{,}799.666$ s, which corresponds to around 2 days of simulated time. Both steady stationary states reached are in agreement with the space physics conditions. After that, the solar wind dataset input starts and the evolution of the solar wind-magnetosphere coupling process is simulated.

This problem is simulated using the HLLD Riemann solver [35] combined with the MinMod limiter under the Courant number $\sigma = 0.4$. The GLM formulation uses the factor $\alpha_p = 0.4$.

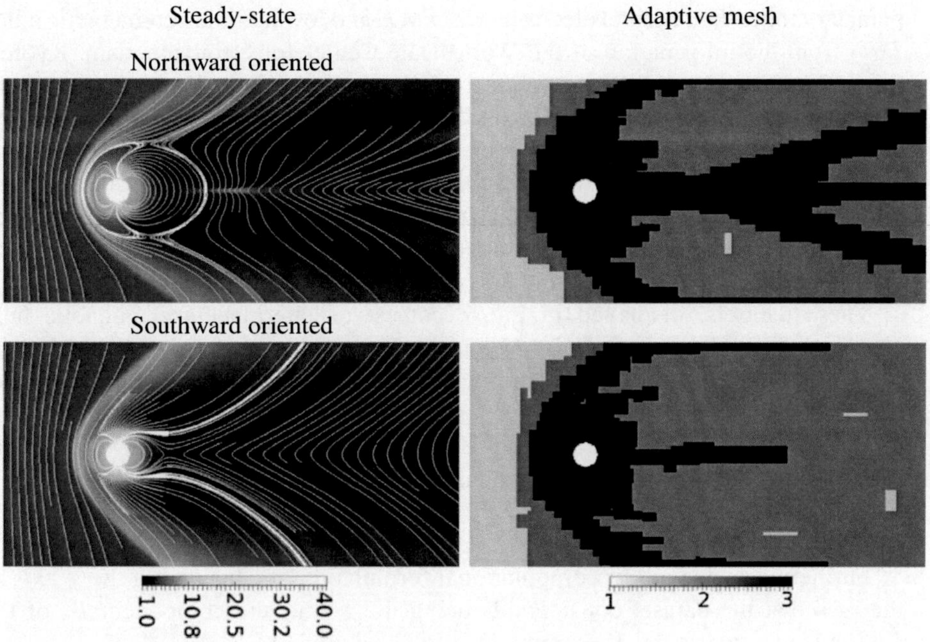

Fig. 5 Steady-state magnetosphere predictions obtained, respectively, from northward- (top) and southward-oriented (bottom) interplanetary magnetic field using an adaptive mesh with finest level corresponding to a 512×256 mesh. Left: density solution (in n/cc) with magnetic field lines. Right: adaptive mesh using the threshold value $\epsilon = 0.05$

Table 9 MAG: errors in pressure p, memory consumption (number of cells and patches used) and CPU time obtained by using several refinement levels L using MinMod limiter at $t_{\text{end}} = 184396$ with a threshold value $\epsilon = 0.05$. The experiments were performed using magnetic field orientation in both north and south directions until the stationary state is obtained

Solar wind	Mesh			Accuracy (p)	Cells		Patches	CPU time	
orientation	Size	Base	L	$\mathbb{L}_1$-error ($\times 10^{-3}$)	#	%	#	Min	%
North	512 × 256	128 × 64	3	3.4158	84,468	65	313	161	74
	512 × 256	256 × 128	2	2.6703	81,784	62	323	159	67
	1024 × 512	256 × 128	3	2.0036	191,176	36	505	849	74
South	512 × 256	128 × 64	3	1.4590	78,556	60	266	93	66
	512 × 256	256 × 128	2	1.1366	74,624	57	257	84	60
	1024 × 512	256 × 128	3	1.1953	167,228	32	416	621	36
	1024 × 512	512 × 256	2	0.9073	189,490	36	293	740	43

Solar Wind Experiments in the Inflow

After the establishment of the bow shock and the magnetosphere at the stationary state, we introduce the interplanetary parameters obtained from the OMNI web service, NASA, corresponding to the period from January 16th to 18th of 2018. By the examination of the geomagnetic disturbance indices (using, for instance, planetary index Kp, auroral electrojet index AE and low latitude geomagnetic index Dst) from the information of the World Data Center for Geomagnetism, Kyoto,[3] this interval corresponds to a typical period of geomagnetically quiet conditions (details in the Appendix). Nevertheless, it still represents a situation of interesting variation in the IMF orientation, which challenges the actual MHD modelling. We use a smoothed version of the data that preserves the primary realistic features, plotted in Fig. 6, to avoid unnecessary oscillations in the simulation. In detail, the filtering process is built with an orthogonal wavelet transform reconstruction with Daubechies family 8 removing the first 6 levels of the wavelet coefficients.

These datasets, displayed in Fig. 6, consist of interplanetary magnetic field components (in Geocentric Solar Magnetospheric reference system, GSM [42, p. 536]) Bx and Bz in nanotesla, the plasma flow speed in kilometre per second, the numerical density in particle per cubic centimetre and solar wind temperature in Kelvin. All these quantities are given in a function of the time in day of year (DOY). Also, the pressure value is obtained using the particle density data n and the temperature T as $p = nkT$, where $k = 1.38064852 \cdot 10^{-23}$ $\mathrm{m^2 kg\, s^{-2} K^{-1}}$ is the Boltzmann constant.

Furthermore, due to our computational formulation of considering the y-axis as the axis that the dataset considers as being the z-axis, the component B_z of the dataset is inputted as the B_y component of the solar wind.

These smoothed time series are introduced as an inflow boundary condition of the northward orientated steady-state magnetosphere, that is used as an approximation of the Earth magnetosphere on Jan 16th 00 : 00h-UT (shown at the top-left panel in Fig. 5). Selected from the numerical evolution as an example, Fig. 7 presents the configuration of the Earth magnetosphere on the instants Jan 17th at 07 : 04h-UT, Jan 17th at 08 : 00h-UT and Jan 17th at 10 : 04h-UT, which, respectively, refer to the effect of the IMF orientation. The upper panel in the figure corresponds to the northward-oriented B_z interval (letter A in Fig. 6), the intermediate one corresponds to the start of negative value (letter B) and the lower panel corresponds to the peak of the southward-oriented interval (letter C). As expected, reconnection processes are noticed in a remarkable way, at the lower panel, in the frontal face and inside the tail of the magnetosphere. The analysis of the dynamic evolution represented in the plots can justify the importance of these kinds of simulation even for geomagnetically quiet conditions on the ground. Interestingly, related to letter C, a time-coincident small intensity effect was noted in the auroral index, and no effect stands out in the other equatorial index (condition reported in the Fig. A.1 in Appendix). A more complete physical investigation is out of the scope of this work. In the simulation figure, the grey pattern used for the density allows clear identification of the interplanetary region, the bow shock, the interplanetary magnetosheath and the own magnetosphere.

[3]World Data Center for Geomagnetism: http://wdc.kugi.kyoto-u.ac.jp/wdc/Sec3.html.

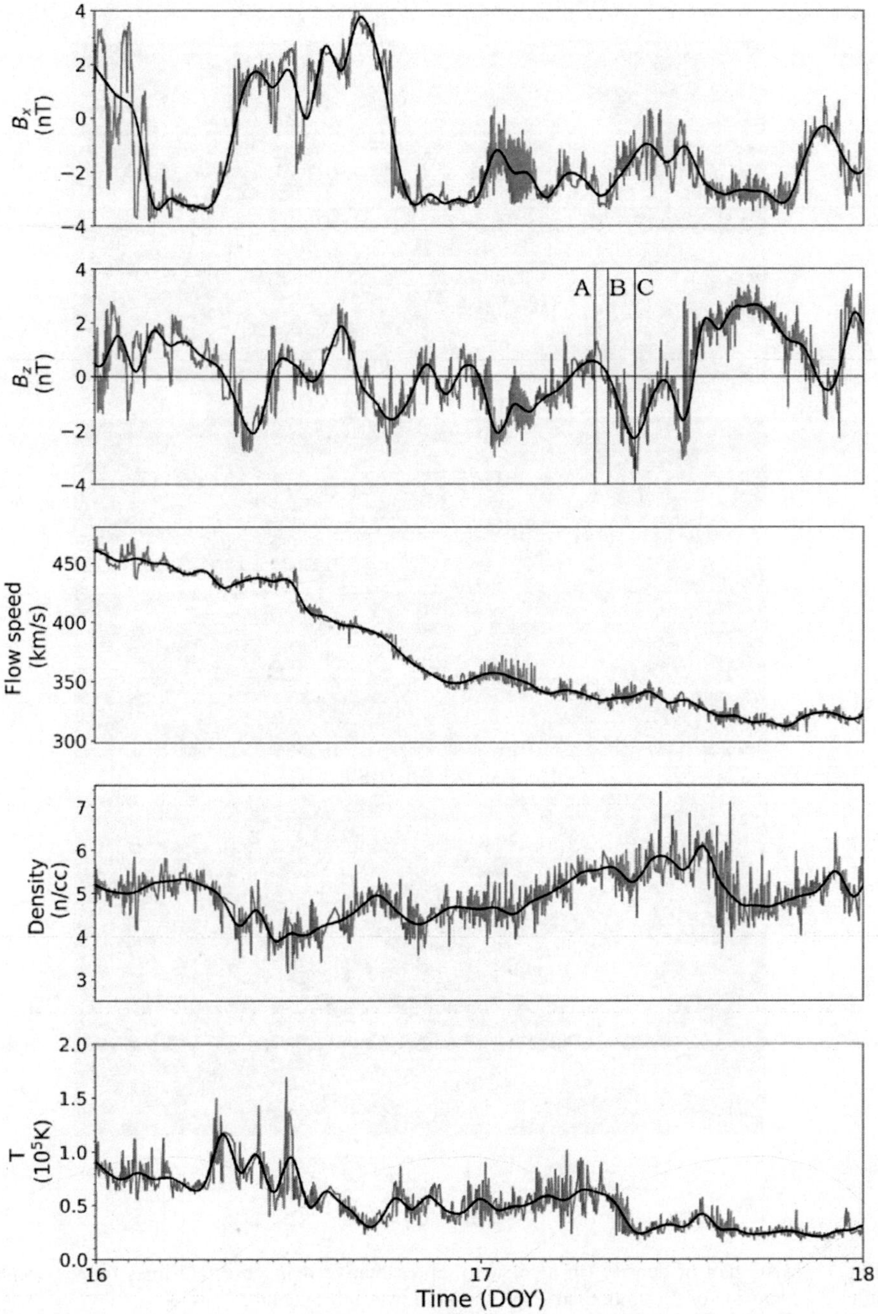

Fig. 6 MAG solar wind inflow conditions corresponding to the period between January 16th and 18th of 2018. The dataset is presented in grey, while the smoother version effectively used on the simulation is presented in black

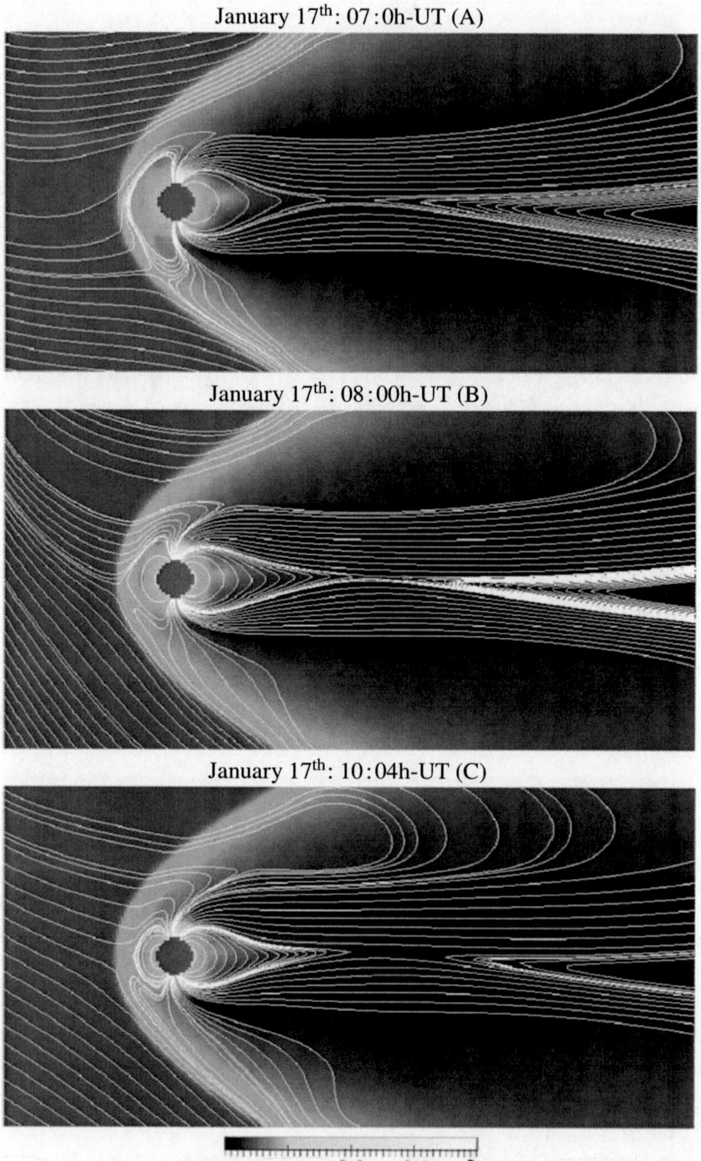

Fig. 7 MAG plot of density (in n/cc) with representative magnetic field lines for the scenarios after the inclusion of the smoothed solar wind inflow satellite data

5 Conclusion

The framework AMROC with its implemented SAMR algorithm, MPI parallelisation, multiresolution criterion and clustering structure was the base of the presented ideal and resistive GLM-MHD solvers with parabolic–hyperbolic magnetic divergence-free correction. These solvers achieved a milestone in the development of an Earth's magnetosphere model. In this work, three physical and numerical challenging tests demonstrate the computational efficiency and memory utilisation of the framework for these new solvers with parallel mesh adaptation, compared to the correspondent solvers using a uniform mesh.

As expected, the performance gain of the MR adaptive MHD simulations depends on the local regularity of the solution in all physical quantities. For instance, excellent adaptive performance can be observed when the solution requires a high-resolution discretisation and has few localised disturbances relative to its domain size. Furthermore, we successfully have validated the proposed solver for a real-world magnetosphere scenario of space plasma, performing a challenging two-dimensional test for the Earth magnetosphere, and additionally, a three-dimensional implementation is also already a work in progress. Nevertheless, from our analyses, further performance enhancements also seem feasible through better parameter choices. In summary, our MHD solvers can deal with complex problems in space research; for instance, as it is, we already have a potentially straightforward numerical space weather forecast model that forthcoming projects will complete.

This work contributes to the rapidly developing area of space weather forecast and is concerned with the inherent computational challenges of reducing memory and CPU time, which—in the three-dimensional case—is still urgent work in process, even for supercomputers.

Acknowledgments The authors thank the FAPESP (grants: 2018/03039-9, 2015/ 25624-2), CNPq (grants: 424352/2018-4, 302226/2018-4, 307083/2017-9, 306038/2015-3) and FINEP (grant: 0112052700) for financial support of this research. MML thanks CNPq (grant: 140626/2014-0) for his doctorate and CAPES (grant: 88882.463276/2019-01) for his post-doctorate scholarship. We are indebted to Eng. V. E. Menconi for his invaluable computational assistance, to M. Sierra-Lorenzo and A. K. F. Gomes for the fruitful discussions, and to Prof. Ogino for the MHD code and scientific discussions that inspired our magnetosphere implementation. We also thank the teams of World Data Center for Geomagnetism, Kyoto, for the geomagnetic indices dataset, and the OMNI web service, NASA, for access to the interplanetary dataset.

We also thank the anonymous reviewers for their comments and suggestions that have improved the final version of this manuscript.

Appendix

Code Organisation

In the context of this work, the AMROC framework, as described in [15] and published online[1] is divided into two main folders: the implementation and compilation folders. The folder *vtf/amroc/amr* contains the base algorithm for a numerical simulation using SAMR methods for a generic system of hyperbolic equations. The files contained in this folder specify the data structures and routines outside the scope of the simulated equations, such as mesh adaptation, mesh distribution per processor, boundary conditions, restriction and prolongation operators, etc. In particular, the function *IntegrateLevel()* in the file *AMRSolver.h* calls the numerical scheme associated with the simulated equation, implemented in the base module, using the *mpass* counter. For each iteration of this counter, the scheme defined in the base is computed and then the ghost cells are updated. Considering the implemented MHD solver, this counter performs three iterations, corresponding to the first Runge–Kutta (RK) step, the second RK step and the divergence cleaning step, respectively. The GLM implementation files are located in the *mhd* directory of the implementation folder. They contain the base virtual functions to perform a generic simulation of the MHD equations for two and three dimensions.

In special, these files contain the time evolution function *Step()*, called from the Generic SAMR solver, and the virtual functions called from this function. The use of virtual functions allows the definition of base functions that may be used for most of the experiments, while allowing the redefinition of these functions in the specific MHD module, if required by the studied problem. In general, the functions from the base module implement numerical operations that are independent from the problem simulated, such as flux computations, limiters and divergence cleaning routines. The problem-specific file located in the respective source directory *src* implements functions that are particular to each experiment. In general, this file contains initial conditions and resistivity and gravity fields. However, if necessary for the experiment, this file may contain redefinitions of virtual functions implemented in the base module. We also have for each simulation an input parameter namelist called solver.in.

Finally, the MHD module in AMROC runs scripts that already contain the commands to convert the output HDF (Hierarchical Data Format) files into binary VTK (Visualization ToolKit) files used for data visualisation in tools such as VisIt[2][10] and ParaView[3][1].

Geomagnetic Disturbances

To attend the interests of the geophysical community, devices to measure the geomagnetic field, designated in a general sense as magnetometers, have been

[1] AMROC webpage: http://www.vtf.website/asc/wiki/bin/view/Amroc.

[2] Visit webpage: https://wci.llnl.gov/simulation/computer-codes/visit/downloads.

[3] Paraview webpage: https://www.paraview.org/download/.

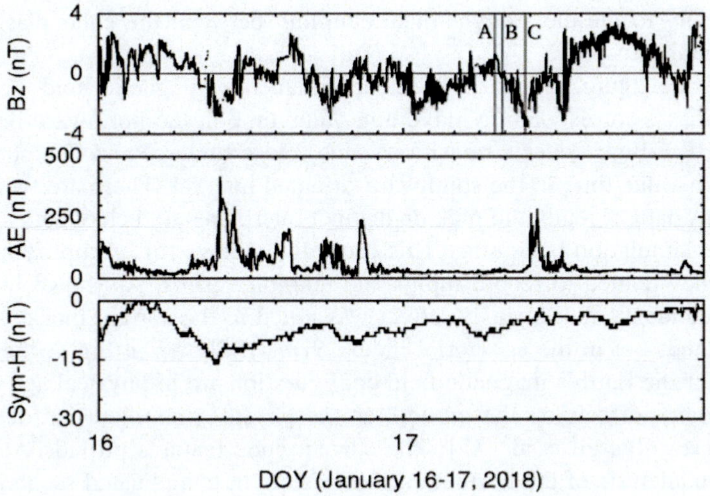

Fig. A.1 MAG: Geomagnetic disturbances shown by the geomagnetic indices: Auroral Electroject (AE) and Symmetric Equatorial ring current effect (Sym-H) corresponding to the period between January 16th and 18th of 2018. Indicated in the interplanetary magnetic field Bz. Letter A indicates northward-oriented field interval, B a transition value and C southward-oriented interval

installed on the ground, nowadays composing a large net spread around the world. One can find more information and specific documentation in the World Data Center for Geomagnetism, Kyoto.[4] Also, related fundamentals on space physics are available in [42]. To quantify the level of geomagnetic disturbance occurring upon the Earth, the interested reader can survey and examine some geomagnetic disturbance indices, for instance, the index Kp for an estimated planetary disturbance behaviour, the index AE for auroral electrojet disturbance effects and the index Dst for a low latitude magnetic disturbance. In our case, we choose and present in Fig. A.1 the interplanetary magnetic field Bz, the index AE and the index Sym-H. This information can be collected effortlessly from the OMNI web service, NASA.[5] Bz is the primary variable responsible for triggering of the magnetic reconnection process (merging of the interplanetary magnetic field lines with the geomagnetic field lines), when this IMF component is a predominantly southward-oriented field (i.e. in opposition to the geomagnetic field orientation), in the frontal side, i.e. towards to the Sun, of the Earth's magnetosphere. AE is the geomagnetic index concerning the modification of the auroral electrojet currents that produce magnetic disturbances in the higher latitudes. *Sym-H* is the index concerning the intensification of an equatorial, symmetrical ring electrical current (at a distance about 6–7 Earth radii) that produces magnetic disturbances in the lower latitudes. Recorded by geomagnetic indices, any geomagnetic variations link

[4]World Data Center for Geomagnetism: http://wdc.kugi.kyoto-u.ac.jp/wdc/Sec3.html.

[5]OMNI web service, NASA: https://omniweb.gsfc.nasa.gov/form/omni_min.html.

intrinsically to the electrodynamical coupling between the solar plasma and the Earth's magnetosphere.

From the figure, indicated in the interplanetary magnetic field Bz, the letter A identifies a corresponding maximum-value time in the northward-oriented field interval, B a time under a transition value (close to zero) and C a corresponding minimum-value time in the southward-oriented interval. There are two reasons to select this dataset region: to pick up distinct interplanetary behaviours and to be far from the simulation beginning. This procedure allows for exemplifying evolution consistency related to record inputs and tangible results. As shown in the plot, a time-coincident small intensity effect was noted in the auroral index, AE, and no effect stands out in the equatorial index, Sym-H. Shown in Fig. 7, the simulation results for the Earth's magnetic field configuration are in physical agreement with the magnetic effects on the ground, as the physics presented and discussed, for instance, by Russell et al. [42]. The current code features provide the means for evolution analysis of the Earth's magnetosphere in complicated scenarios, such as investigations for geomagnetically quiet conditions.

References

1. Ahrens, J., Geveci, B., Law, C.: ParaView. Los Alamos National Laboratory, Los Alamos (2005)
2. Bell, J., Berger, M.J., Saltzmann, J., Welcome, M.: Three-dimensional adaptive mesh refinement for hyperbolic conservation laws. SIAM J. Sci. Comput. **15**, 127–138 (1994)
3. Berger, M.J.: Adaptive mesh refinement for hyperbolic partial differential equations. Ph.D. Thesis, Stanford University, Stanford (1982)
4. Berger, M.J., Colella, P.: Local adaptive mesh refinement for shock hydrodynamics. J. Comput. Phys. **82**(1), 64–84 (1989)
5. Berger, M.J., Oliger, J.: Adaptive mesh refinement for hyperbolic partial differential equations. J. Comput. Phys. **53**, 484–512 (1984)
6. Bittencourt, J.A.: Fundamentals of Plasma Physics, 3rd edn. Springer, New York (2004)
7. Brackbill, J.U., Barnes, D.C.: The effect of nonzero $\nabla \cdot B$ on the numerical solution of the magnetohydrodynamic equations. J. Comput. Phys. **35**(3), 426–430 (1980)
8. Chapman, S., Ferraro, V.C.A.: A new theory of magnetic storms. Nature **126**(3169), 129–130 (1930)
9. Chapman, S., Ferraro, V.C.A.: A new theory of magnetic storms. Terr. Magn. Atmos. Electr. **36**, 171–186 (1931)
10. Childs, H., Brugger, E., Whitlock, B., Meredith, J., Ahern, S., Pugmire, D., Biagas, K., Miller, M., Harrison, C., Weber, G.H., Krishnan, H., Fogal, T., Sanderson, A., Garth, C., Bethel, E.W., Camp, D., Rübel, O., Durant, M., Favre, J.M., Navrátil, P.: VisIt: an end-user tool for visualizing and analyzing very large data. In: Bethel, E.W., Childs, H., Hansen, C. (eds.) High Performance Visualization–Enabling Extreme-Scale Scientific Insight, 1st edn., Chapter 16, pp. 357–372. Chapman and Hall/CRC, New York (2012)
11. Cohen, A.: Wavelet methods in numerical analysis. In: Ciarlet, P.G., Lions, J.L. (eds.), Handbook of Numerical Analysis, vol. VII. Elsevier, Amsterdam (2000)
12. Dedner, A., Kemm, F., Kröner, D., Munz, C.-D., Schnitzer, T., Wesenberg, M.: Hyperbolic divergence cleaning for the MHD equations. J. Comput. Phys. **175**(2), 645–673 (2002)
13. Deiterding, R.: Parallel adaptive simulation of multi-dimensional detonation structures. Ph.D. Thesis, Brandenburgische Technische Universität Cottbus (2003)

14. Deiterding, R.: Construction and application of an AMR algorithm for distributed memory computers. In: Plewa, T., Linde, T., Weirs, V.G. (eds.), Adaptive Mesh Refinement – Theory and Applications, pp. 361–372. Springer, Berlin (2005)
15. Deiterding, R.: Block-structured adaptive mesh refinement – theory, implementation and application. ESAIM: Proc. **34**, 97–150 (2011)
16. Deiterding, R., Domingues, M.O.: Evaluation of multiresolution mesh adaptation criteria in the AMROC framework. In: Proceedings of the Fifth International Conference on Parallel, Distributed, Grid and Cloud Computing for Engineering, vol. 111. Civil Comp Press (2017)
17. Deiterding, R., Domingues, M.O., Gomes, S.M., Roussel, O., Schneider, K.: Adaptive multiresolution or adaptive mesh refinement: a case study for 2D Euler equations. ESAIM Proc. **29**, 28–42 (2009)
18. Deiterding, R., Domingues, M.O., Schneider, K.: Multiresolution analysis as a criterion for effective dynamic mesh adaptation – a case study for Euler equations in the SAMR framework AMROC. Comput. Fluids **205**, 104583 (2020)
19. Domingues, M.O., Deiterding, R., Moreira Lopes, M., Gomes, A.K.F., Mendes, O., Schneider, K.: Wavelet-based parallel dynamic mesh adaptation for magnetohydrodynamics in the AMROC framework. Comput. Fluids **190**, 374–381 (2019)
20. Domingues, M.O., Gomes, S.M., Roussel, O., Schneider, K.: Adaptive multiresolution methods. ESAIM Proc. **34**, 1–96 (2011)
21. Eastwood, J.P., Biffis, E., Hapgood, M.A., Green, L., Bisi, M.M., Bentley, R.D., Wicks, R., McKinnell, L.A., Gibbs, M., Burnett, C.: The economic impact of space weather: where do we stand? Risk Anal. **37**(2) (2017)
22. Feng, X.: Magnetohydrodynamic Modeling of the Solar Corona and Heliosphere. Springer, Singapore (2020)
23. Fromang, S., Hennebelle, P., Teyssier, R.: A high order Godunov scheme with constrained transport and adaptive mesh refinement for astrophysical MHD. Astron. Astrophys. **457**(2), 371–384 (2006)
24. Gheller, C.: ENZO and RAMSES codes for computational astrophysics. Technical Report, Swiss National Supercomputing Centre (2017). Available online in https://hpc-forge.cineca.it/files/CoursesDev/public/2017/HPC_methods_for_Computational_Fluid_Dynamics_and_Astrophysics/Bologna/ASTR02-ENZO_and_RAMSES_Codes-Gheller.pdf
25. Goedbloed, P., Keppens, R.: Lecture notes in magnetohydrodynamics of astrophysical plasmas. Chapter 4: The MHD model. Technical report, Utrecht University, Sep-2004/Jan-2005. Available online in https://perswww.kuleuven.be/~u0016541/MHD_sheets_pdf/nsap430m.06.4.pdf
26. Hargreaves, J.K.: The Solar-Terrestrial Environment. Cambridge University Press, Cambridge (1992)
27. Harten, A.: Multiresolution algorithms for the numerical solution of hyperbolic conservation laws. Commun. Pure Appl. Math. **48**(12), 1305–1342 (1995)
28. Harten, A., Lax, P.D., van Leer, B.: On upstream differencing and Godunov-type schemes for hyperbolic conservation laws. SIAM Rev. **25**(1), 35–61 (1983)
29. Hopkins, P.F.: A constrained-gradient method to control divergence errors in numerical MHD. Mon. Not. R. Astron. Soc. **462**, 576–587 (2016)
30. Jiang, R.-L., Fang, C., Chen, P.-F.: A new MHD code with adaptive mesh refinement and parallelization for astrophysics. Comput. Phys. Commun. **183**(8), 1617–1633 (2012)
31. Koren, B.: A robust upwind discretization method for advection, diffusion and source terms. In: Vreugdenhil, C.B., Koren, B. (eds.), Numerical Methods for Advection-Diffusion Problems. Notes on Numerical Fluid Mechanics, pp. 117–138. Vieweg, Germany (1993)
32. Landau, L.D., Lifshitz, E.M., Pitaevskii, L.P.: Electrodynamics of Continuous Media, vol. 8. Course of Theoretical Physics S, Pergamon, 2nd edn. (2004)
33. LeVeque, R.J.: Numerical Methods for Conservation Laws. Birkhäuser Verlag, Basel (1990)
34. Mignone, A., Tzeferacos, P., Bodo, G.: High-order conservative finite difference GLM–MHD schemes for cell-centered MHD. J. Comput. Phys. **229**(17), 5896–5920 (2010)

35. Miyoshi, T., Kusano, K.A.: A multi-state HLL approximate Riemann solver for ideal magnetohydrodynamics. J. Comput. Phys. **208**, 315–344 (2005)
36. Moreira Lopes, M., Deiterding, R., Gomes, A.K.F., Mendes, O., Domingues, M.: An ideal compressible magnetohydrodynamic solver with parallel block-structured adaptive mesh refinement. Comput. Fluids **173**, 293–298 (2018)
37. Müller, S.: Adaptive Multiscale Schemes for Conservation Laws, vol. 27. Lecture Notes in Computational Science and Engineering. Springer, Heidelberg (2003)
38. Ogino, T.: A Three-Dimensional MHD simulation of the interaction of the Solar Wind with the Earth's Magnetosphere: the generation of field-aligned currents. J. Geophys. Res. **91**(A6), 6791–6806 (1986)
39. Ogino, T.: Two-dimensional MHD code. In: Omura, Y., Matsumoto, H. (ed.), Computer Space Plasma Physics: Simulations and Software, pp. 161–191. Terra Scientific Publishing, Tokyo (1993)
40. Ogino, T., Walker, R.J., Ashour-Abdalla, M.: A global magnetohydrodynamic simulation of the magnetosheath and magnetosphere when the interplanetary magnetic field is northward. IEEE Trans. Plasma Sci. **20**(6), 817–828 (1992)
41. Roe, P.L.: Characteristic-based schemes for the Euler equations. Annu. Rev. Fluid Mech. **18**(1), 337–365 (1986)
42. Russell, C.T., Luhmann, J.G., Strangeway, R.J.: Space Physics: An Introduction. Cambridge University Press, Cambridge (2016)
43. Schrijver, C.J., Kauristie, K., Aylward, A.D., Denardini, C.M., Gibson, S.E., Glover, A., Gopalswamy, N., Grande, M., Hapgood, M., Heynderickx, D., Jakowski, N., Kalegaev, V.V., Lapenta, G., Linker, J.A., Liu, S., Mandrini, C.H., Mann, I.R., Nagatsuma, T., Nandy, D., Obara, T., O'Brien, T.P., Onsager, T., Opgenoorth, H.J., Terkildsen, M., Valladares, C.E., Vilmer, N.: Understanding space weather to shield society: A global road map for 2015–2025 commissioned by COSPAR and ILWS. Adv. Space Res. **55**(12), 2745–2807 (2015)
44. Stone, J.M., Gardiner, T.A., Teuben, P., Hawley, J.F., Simon, J.B.: Athena: a new code for astrophysical MHD. Astrophys. J. Suppl. Ser. **178**(1), 137–177 (2008)
45. Toro, E.F.: Riemann Solvers and Numerical Methods for Fluid Dynamics: A Practical Introduction. Springer, Berlin (1999)
46. Tóth, G.: The $\nabla \cdot B$ constraint in shock-capturing magnetohydrodynamics codes. J. Comput. Phys. **161**, 605–652 (2000)
47. Tóth, G., van der Holst, B., Sokolov, I.V., De Zeeuw, D.L., Gombosi, T.I., Fang, F., Manchester, W.B., Meng, X., Najib, D., Powell, K.G., Stout, Q.F., Glocer, A., Ma, Y.-J., Opher, M.: Adaptive numerical algorithms in space weather modeling. J. Comput. Phys. **231**, 870–903 (2012)
48. van Leer, B.: Towards the ultimate conservative difference scheme. II. Monotonicity and conservation combined in a second-order scheme. J. Comput. Phys. **14**(4), 361–370 (1974)
49. van Leer, B.: Towards the ultimate conservative difference scheme III. Upstream-centered finite-difference schemes for ideal compressible flow. J. Comput. Phys. **23**(3), 263–275 (1977)
50. van Leer, B.: Towards the ultimate conservative difference scheme. V. A second-order sequel to Godunov's method. J. Comput. Phys. **32**(1), 101–136 (1979)
51. van Albada, G.D., van Leer, B., Roberts, W.W.: A comparative study of computational methods in cosmic gas dynamics. In: Hussaini, M.Y., van Leer, B., van Rosendale, J. (eds.), Upwind and High-Resolution Schemes, pp. 95–103. Springer, Berlin (1997)

Verification of the WALE Large Eddy Simulation Model for Adaptive Lattice Boltzmann Methods Implemented in the AMROC Framework

Christos Gkoudesnes and Ralf Deiterding

Abstract We detail the verification of the WALE large eddy simulation turbulence model for application in cell-based lattice Boltzmann methods, as implemented in our generic Cartesian structured adaptive mesh refinement framework AMROC. We demonstrate how to effectively apply the test case of decaying homogeneous isotropic turbulence to verify the core WALE implementation against higher resolved direct numerical simulations and the constant-coefficient Smagorinsky turbulence model. Both standard and regularised single relaxation collision models are analysed systematically. While our results confirm the established observation that the standard collision model yields less dissipative energy spectra, novel quantitative evidence is given that this positive behaviour comes at the cost of unphysical perturbations in high wavenumbers. In order to allow unaltered application of the finite-difference stencils intrinsic to the WALE approach in real-world flow situations, a new method is presented for ensuring consistent boundary conditions in microscopic distribution functions as well as in macroscopic variables. The benefit of the proposed technique is shown for dynamically adaptive simulations of flow around a sphere at Reynolds number 1000 and compared to a large eddy simulation using the constant-coefficient Smagorinsky model.

1 Introduction

In recent years, the lattice Boltzmann method (LBM) [19, 31] has achieved stupendous success in a variety of scientific fields. Application examples can be found for instance in [1, 15, 22, 29, 33–35]. Its computationally inexpensive numerical scheme, straightforward parallelisation and close to linear parallel scalability make it a powerful alternative for subsonic flow simulations compared to the mainstream computational fluid dynamics solvers that discretise the Navier–Stokes equations

C. Gkoudesnes (✉) and R. Deiterding
Aerodynamics and Flight Mechanics Research Group, School of Engineering, University of Southampton, Southampton, UK
e-mail: C.Gkoudesnes@soton.ac.uk; R.Deiterding@soton.ac.uk

R. Deiterding et al. (eds.), *Cartesian CFD Methods for Complex Applications*,
SEMA SIMAI Springer Series 3, https://doi.org/10.1007/978-3-030-61761-5_6

and usually employ finite volume schemes. Thanks to a time-explicit numerical update and intrinsically low numerical dissipation, the LBM lends itself particularly to large eddy simulations (LES) of engineering applications involving high Reynolds number flows. The employment of Cartesian meshes, and characteristic for the LBM, in addition, allows easy and automatic mesh generation and, hence, has the potential of reducing the time for setting up a simulation considerably, particularly with complex geometries. However, the drawback of the Cartesian approach is that a significant number of cells usually need to be deployed in the vicinity of the body in order to accurately approximate its shape. In the case of uniform grids, this can lead to prohibitively large meshes. A possibility to mitigate this issue is the extension of the LBM to body-fitted structured [28] or hybrid meshes [9]. The other—more common—approach is the use of levels of Cartesian refinement. This approach can be further optimised by the implementation of solution adaptive mesh refinement (AMR).

The AMROC (Adaptive Mesh Refinement in Object-oriented C++) framework [6] implements patch-based, structured adaptive mesh refinement (SAMR) generically for time-explicit finite volume methods. The LBM has been incorporated into AMROC by formulating it on cell-based data structures; treatment of embedded boundaries with a level-set-based ghost-fluid-type approach allows for an effective handling of moving solid bodies. Examples of successful AMROC-LBM simulations, primarily in the laminar flow regime, can be found for instance in [7, 8, 10, 11, 18, 20]. The present paper reports on verification and validation of a variety of new developments in the AMROC-LBM solver, in particular, the wall-adapting local eddy viscosity (WALE) turbulence model [25] and the newly implemented regularised single relaxation time (SRT) collision operator [21]. In the procedure of applying LES models that are based on finite-difference stencils, a new algorithm for imposing macroscopic variables in ghost cells, after the application of "bounce-back" boundary conditions, is presented and tested here for the very first time. Two validation tests are discussed in detail, namely, the decaying homogeneous isotropic turbulence in a periodic box benchmark and turbulent flow around a sphere at Reynolds number 1000. Comparing the spectra from the STAndard (STA) and REGularised (REG) SRT operators in the former case, useful information will be extracted. The efficiency and performance of the WALE model will also be cross-verified against the Constant SMAgorinsky (CSMA) model.

The paper is organised as follows: In Sect. 2, we present the LBM equations both for the STA and REG SRT operators, the formulas for the CSMA and WALE models and the SAMR strategy as implemented in AMROC. Section 3 details the improved boundary condition implementation in AMROC-LBM, both for the domain and the embedded non-Cartesian surface boundaries, and the new algorithm for imposing microscopic as well as macroscopic variables in ghost cells will be reported. The results of the two validation test cases and their discussion can be found in Sect. 4. Finally, the conclusions are drawn in Sect. 5.

2 Methodology

In this section, we review the lattice Boltzmann method and the newly implemented REG SRT collision model. Moreover, the formulas describing the two LES models, employed in this paper, are introduced. Finally, we summarise the SAMR strategy that is applied in the AMROC-LBM solver.

2.1 Lattice Boltzmann Method

The discrete lattice Boltzmann equation, describing the evolution of the distribution functions f_α with the SRT collision model and without an external force, reads

$$\partial_t f_\alpha + \mathbf{e}_\alpha \cdot \nabla f_\alpha = \tau^{-1}(f_\alpha^{\mathrm{eq}} - f_\alpha), \tag{1}$$

where τ is the discrete relaxation time. We chose the standard discretisation in space and time based on a finite-difference scheme and a two-step procedure. The first operation that is applied during the time step update is the streaming

$$\check{f}_\alpha(\mathbf{x} + \mathbf{e}_\alpha \Delta t, t) = f_\alpha(\mathbf{x}, t), \tag{2}$$

where $\check{f}_\alpha$ is the intermediate value of the distribution function between the two steps. The second operation is the collision. For the STA SRT model, it is defined as

$$f_\alpha(\mathbf{x}, t + \Delta t) = \check{f}_\alpha(\mathbf{x}, t) + \frac{\Delta t}{\tau}(f_\alpha^{\mathrm{eq}}(\mathbf{x}, t) - \check{f}_\alpha(\mathbf{x}, t)). \tag{3}$$

The discrete relaxation time τ in LBM is given as

$$\tau = \frac{\nu + \Delta t c_s^2/2}{c_s^2}, \tag{4}$$

where ν is the kinematic viscosity and c_s is the physical speed of sound of the fluid. The number of the lattice velocities $\mathbf{e}_\alpha$ depends on the employed LBM model. In the current research work, the D3Q19 model was used, with the 19 lattice directions defined as

$$\mathbf{e}_\alpha = \begin{cases} 0, & w_\alpha = \frac{12}{36}, \quad \alpha = 0, \\ (\pm 1, 0, 0)c, (0, \pm 1, 0)c, (0, 0, \pm 1)c, & w_\alpha = \frac{2}{36}, \quad \alpha = 1, \ldots, 6, \\ (\pm 1, \pm 1, 0)c, (\pm 1, 0, \pm 1)c, (0, \pm 1, \pm 1)c, & w_\alpha = \frac{1}{36}, \quad \alpha = 7, \ldots, 18. \end{cases} \tag{5}$$

The notation c is the ratio $\Delta x/\Delta t$. The Maxwellian equilibrium distribution function is truncated to second order, yielding

$$f_\alpha^{\text{eq}}(\mathbf{x},t) = w_\alpha\,\rho\left[1+\frac{\mathbf{e}_\alpha\cdot\mathbf{u}}{c_s^2}+\frac{(\mathbf{e}_\alpha\cdot\mathbf{u})^2}{2c_s^4}-\frac{\mathbf{u}\cdot\mathbf{u}}{2c_s^2}\right], \tag{6}$$

with the user option in AMROC-LBM to increase the latter approximation to third order for slightly improved accuracy and stability. The macroscopic variables, density ρ, velocity vector $\mathbf{u}$ and pressure p, can be evaluated through the first two moments of the distribution functions f_α as

$$\rho(\mathbf{x},t) = \sum_\alpha f_\alpha(\mathbf{x},t), \tag{7a}$$

$$\rho(\mathbf{x},t)u_i(\mathbf{x},t) = \sum_\alpha e_{\alpha i} f_\alpha(\mathbf{x},t), \tag{7b}$$

$$p(\mathbf{x},t) = \rho(\mathbf{x},t)\,c_s^2. \tag{7c}$$

At this point, it is helpful to introduce the non-equilibrium part of the distribution functions, $f_\alpha^{\text{neq}}(\mathbf{x},t) = f_\alpha(\mathbf{x},t) - f_\alpha^{\text{eq}}(\mathbf{x},t)$, and, by utilisation of the second moment, to obtain the momentum flux tensor Π_{ij}^{neq} as

$$\Pi_{ij}^{\text{neq}} = \sum_\alpha e_{\alpha i} e_{\alpha j} f_\alpha^{\text{neq}}(\mathbf{x},t). \tag{8}$$

This tensor is analogous to the strain rate in the Navier–Stokes equations and will be useful for the subsequent discussion.

2.1.1 Regularised Single Relaxation Time Collision Model

To further improve the stability of the AMROC-LBM solver for high Reynolds number flows, a second collision model, namely REG SRT, proposed by Latt and Chopard [21], has been recently implemented. The idea is to *regularise* the non-equilibrium part of the distribution functions before one applies the collision step. This procedure reads

$$f_\alpha^{(1)}(\mathbf{x},t) = \frac{w_\alpha}{2c_s^4} Q_{\alpha ij}\Pi_{ij}^{\text{neq}}, \tag{9}$$

where $Q_{\alpha ij} = e_{\alpha i}e_{\alpha j} - c_s^2\delta_{ij}$ and Π_{ij}^{neq} is estimated from Eq. (8). In this way, the non-equilibrium part retains the symmetry that is imposed by its relation with the viscous stress tensor and the strain rate. This extra step transforms the computation of the collision operation, Eq. (3), to

$$f_\alpha^{\text{reg}}(\mathbf{x}, t + \Delta t) = f_\alpha^{\text{eq}}(\mathbf{x}, t) + (1 - \frac{\Delta t}{\tau})\check{f}_\alpha^{(1)}(\mathbf{x}, t). \tag{10}$$

2.2 *Large Eddy Simulation*

In AMROC-LBM, the integration of the LES models in the solver is based on the eddy viscosity approach [16]. In the case of a direct numerical simulation (DNS) with the LBM, the discrete relaxation time Eq. (4) is a global variable depending only on the physical speed of sound, the viscosity of the gas and the time step. The idea is that altering the relaxation time is analogous to changing the mean free path of the particles. Invoking the mixing length theory of Prandtl, one can argue that altering the mean free path is equivalent to changing the viscosity, leading to the idea of a turbulent eddy viscosity ν_t. The general formula to calculate the eddy viscosity is

$$\nu_t = (C\Delta x)^2 OP_{\text{LES}}, \tag{11}$$

where C is a constant and OP_{LES} is a time scale estimated differently by each LES model. Therefore, in the case of an LBM LES, the physical viscosity is replaced by an effective viscosity $\nu^\star = \nu + \nu_t$. This alteration also affects the calculation of the discrete relaxation time τ that is replaced by an effective value $\tau^\star$.

2.2.1 Constant Smagorinsky Model

For the case of the CSMA model [30], the eddy viscosity is computed as

$$\nu_t = (C_S\Delta x)^2|\overline{S}|, \tag{12}$$

where $|\overline{S}| = \sqrt{2\overline{S}_{ij}\overline{S}_{ij}}$ is the intensity of the filtered strain rate. The constant C_S is a user parameter, with a usual value between 0.1 and 0.2. As already mentioned in Sect. 2.1.1, the non-equilibrium part of the distribution function f_α^{neq} can be used to estimate the strain rate locally per cell as

$$\overline{S}_{ij} = -\frac{1}{2\rho c_s^2\tau^\star}\Pi_{ij}^{\text{neq}}. \tag{13}$$

One can substitute Eq. (13) in Eq. (12), and after some algebra, an explicit equation for the effective discrete relaxation time $\tau^{\star}$ in each cell is retrieved:

$$\tau^{\star} = \frac{\tau}{2} + \sqrt{\frac{\tau^2}{4} + \frac{C_S^2 \Delta x^2 |\overline{\Pi}_{ij}^{\mathrm{neq}}|}{2\rho c_s^4}}, \tag{14}$$

with τ computed by Eq. (4).

2.2.2 Wall-Adapting Local Eddy Viscosity Model

The idea of the WALE model is to employ a more advanced operator for the characteristic time scale OP_{LES} that can effectively reduce the eddy viscosity to zero at the wall, thereby reproducing the proper scaling $\nu_t \sim y^{+^3}$ without the need for a damping function or any other location-dependent strategy [25]. The new operator is a function of both the strain rate S_{ij} and the rotation rate Ω_{ij}. It reads

$$OP_{\mathrm{WALE}} = \frac{(\mathcal{J}_{ij}\mathcal{J}_{ij})^{\frac{3}{2}}}{(\overline{S}_{ij}\overline{S}_{ij})^{\frac{5}{2}} + (\mathcal{J}_{ij}\mathcal{J}_{ij})^{\frac{5}{4}}}, \tag{15}$$

where $\mathcal{J}_{ij}$ is

$$\mathcal{J}_{ij} = \overline{S}_{ik}\overline{S}_{kj} + \overline{\Omega}_{ik}\overline{\Omega}_{kj} - \frac{1}{3}\delta_{ij}(\overline{S}_{mn}\overline{S}_{mn} - \overline{\Omega}_{mn}\overline{\Omega}_{mn}). \tag{16}$$

Therefore, in this case, the eddy viscosity is calculated as

$$\nu_t = (C_w \Delta x)^2 OP_{\mathrm{WALE}}, \tag{17}$$

where C_w is the constant of the model and is equal to 0.5.

Compared to the CSMA model, which retains the locality of the collision step, the WALE model is based on central finite differences for estimating the rotation and strain rates. This stencil operation destroys the locality of the WALE collision operation, adds extra computational burden and, as we will present below, requires special attention in the application of some boundary conditions.

2.3 *Structured Dynamic Mesh Adaptation*

The AMROC framework provides the capability of dynamic mesh adaptation, utilising user-defined refinement indicators on fully parallelised meshes. Its AMR strategy is based on the block-structured and recursive adaptive mesh refinement method for hyperbolic conservation laws after Berger and Collela [2]. By formulating the LBM on cell-based data structures, the method can be made to fit smoothly into the SAMR execution procedure. A positive side effect of the cell-based formulation is that the scheme becomes conservative in ρ and $\rho\mathbf{u}$.

In the SAMR approach, finite volume cells are clustered with a special algorithm into non-overlapping rectangular grids. The grids have a suitable layer of halo cells for synchronisation and applying inter-level and physical boundary conditions. Refinement levels are integrated recursively and by updating the sequence of grids on each level in a for loop using the same numerical update routine. The spatial mesh width Δx_l and the time step Δt_l on level l are refined by the same factor r_l, where we assume $r_l \geq 2$ for $l > 0$ and $r_0 = 1$. In order to ensure that the same gas, with identical speed of sound and kinematic viscosity, is approximated on all levels of the SAMR hierarchy, with the alteration of Δx_l and thus Δt_l, the discrete relaxation time τ cannot be constant but needs to be adjusted according to Eq. (4) for the update on each level. In addition to this, the interface region requires a specialised treatment. Distinguishing between the streaming $\mathcal{S}$, Eq. (2), and collision $\mathcal{C}$, Eq. (3), the crucial steps of our method are as follows:

1. Use coarse grid distributions $f^C_{\alpha,\text{in}}$ that propagate into the fine grid, cf. Fig. 1a, to construct initial fine grid halo values $f^f_{\alpha,\text{in}}$ by interpolation (Fig. 1b, top).
2. Stream $\check{f}^f_\alpha := \mathcal{S}(f^f_\alpha)$ on entire fine mesh. Collision $f^f_\alpha := \mathcal{C}(\check{f}^f_\alpha)$ is applied only in the interior cells (yellow in Fig. 1b, top). Repeat $r_l - 1$ times.

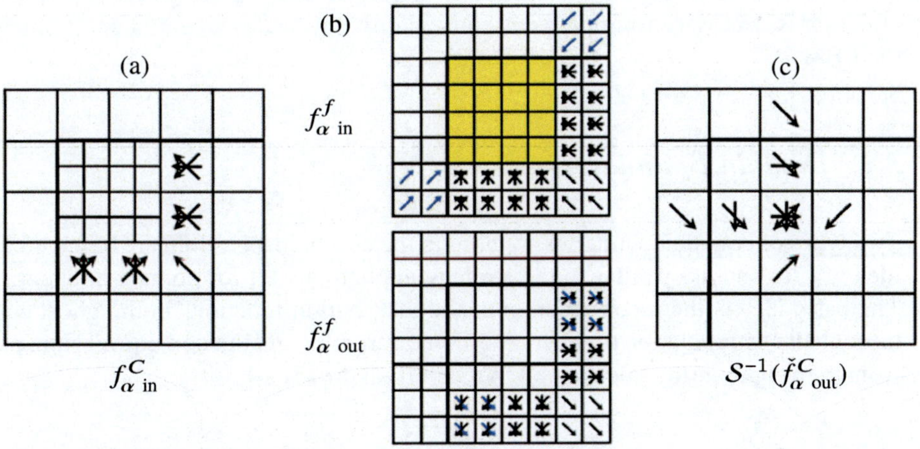

Fig. 1 Visualisation of distributions involved in necessary data exchange at a coarse-fine boundary. The thick black lines indicate a physical boundary. (**a**) Coarse distributions going into fine grid; (**b**) ingoing interpolated fine distributions in halos (top), outgoing distributions in halos after two fine-level transport steps (bottom); (**c**) averaged distributions replacing coarse values before update is repeated in cells next to boundary

3. Average outgoing distributions from fine grid halos (Fig. 1b, bottom) to obtain $\check{f}^{C}_{\alpha,\text{out}}$.
4. Reverse streaming for averaged outgoing distributions, $\tilde{f}^{C}_{\alpha,\text{out}} := \mathcal{S}^{-1}(\check{f}^{C}_{\alpha,\text{out}})$, and overwrite those in the previous coarse grid time step, cf. Fig. 1c.
5. Repeat LBM update on coarse grid cells next to coarse-fine boundary only.

This algorithm is computationally equivalent to the method by Chen et al. [4] but tailored to the SAMR recursion that updates coarse grids in their entirety before fine grids are computed. Because of the nonlinearity of the collision operator C, it becomes necessary under this paradigm to repeat the LBM update for those coarse grid cells that share a face or corner with a fine grid. A comprehensive verification of the adaptive method in AMROC-LBM can be found in [11].

3 Boundary Conditions in AMROC-LBM

The utilisation of halo or ghost cells is intrinsic to the SAMR approach. In order to achieve an efficient and parallel update of the subgrids on each level of the hierarchy, it is of crucial importance to synchronise and apply boundary conditions in ghost cells before the execution of the numerical update routine. Similarly, the modification of internal cells in order to realise geometrically complex embedded wall boundary conditions is in AMROC-LBM equally carried out immediately before the LBM update. While the routines for synchronisation and first-order accurate inter-level boundary conditions in AMROC are generic, cf. [6], special attention is necessary to implement high-quality physical boundary conditions for the LBM.

3.1 Domain Boundaries

Two types of domain boundaries are used in the current paper. Both are based on the idea of [36] and are similar to the strategy applied in [32] for domain boundaries. The basic idea is the reconstruction of the distribution functions in the ghost cells through the extrapolation of the macroscopic variables and the non-equilibrium part from the neighbouring interior cell in the normal direction.

3.1.1 Inlet

In the case of an inlet boundary, the vector of velocities is imposed and the density $\rho(\mathbf{x}_{\text{BC}})$ in the boundary ghost cell is unknown. Assuming a zero gradient, the density of the first neighbouring normal interior fluid cell $\rho(\mathbf{x}_{\text{F}})$ is extrapolated as

$$\rho(\mathbf{x}_{\text{BC}}) = \rho(\mathbf{x}_{\text{F}}). \tag{18}$$

Having obtained the density, one can proceed with estimating the equilibrium part $f_\alpha^{\mathrm{eq}}(\mathbf{x}_{\mathrm{BC}})$ of the distribution function by applying Eq. (6). The same extrapolation can be used to estimate a value for the non-equilibrium part as

$$f_\alpha^{\mathrm{neq}}(\mathbf{x}_{\mathrm{BC}}) = f_\alpha(\mathbf{x}_{\mathrm{F}}) - f_\alpha^{\mathrm{eq}}(\mathbf{x}_{\mathrm{F}}). \tag{19}$$

With these two values, one can now reconstruct the distribution function in the boundary ghost cell as

$$f_\alpha(\mathbf{x}_{\mathrm{BC}}) = f_\alpha^{\mathrm{eq}}(\mathbf{x}_{\mathrm{BC}}) + f_\alpha^{\mathrm{neq}}(\mathbf{x}_{\mathrm{BC}}). \tag{20}$$

3.1.2 Outlet

In the case of the outlet boundary, the density $\rho(\mathbf{x}_{\mathrm{BC}})$ is imposed and the velocity field needs to be extrapolated from the normal neighbouring interior fluid cell. Following the same methodology as in case of the inlet boundary, the extrapolation formula is

$$\mathbf{u}(\mathbf{x}_{\mathrm{BC}}) = \mathbf{u}(\mathbf{x}_{\mathrm{F}}). \tag{21}$$

Having obtained both density and velocity in the ghost cell location, the equilibrium functions, Eq. (6), can be calculated. The non-equilibrium part is also extrapolated following Eq. (19). Finally, the reconstruction of $f_\alpha(\mathbf{x}_{\mathrm{BC}})$ is according to Eq. (20).

3.2 Embedded Wall Boundaries

In the AMROC software, non-Cartesian boundaries are represented implicitly on the adaptive Cartesian grid by utilising a scalar level set function φ that stores the distance to the boundary surface. The boundary surface is located exactly at $\varphi = 0$, and the boundary outer normal in every mesh point can be evaluated as $\mathbf{n} = -\nabla\varphi/|\nabla\varphi|$ [5]. A fluid cell is treated as an embedded ghost cell if its midpoint satisfies $\varphi < 0$.

Real-world geometries are considered in AMROC as triangular surface meshes, cf. [6]. The computation of the level set distance information in every Cartesian cell midpoint could principally be accomplished by simply iterating over the entire surface mesh; yet, this would lead to detrimental performance for increasing problem size. Instead, we employ a specially developed algorithm based on characteristic reconstruction and scan conversion by Mauch [23] that is used to compute the distance exactly only in a small band around the embedded structure.

For imposing no-slip wall boundaries on the LBM, we choose in this paper the bounce-back algorithm of [3]. The idea of this methodology is to enhance the standard half-way bounce-back scheme with a spatial interpolation of first- or second-order accuracy to handle curved boundaries. The interpolation weight q is the ratio between the distance to the wall from the first fluid cell to the grid spacing. Based on the value of q and in case of the first-order accurate interpolation, there are two possibilities:

$$f_{\mathrm{opp}(\alpha)}(\mathbf{x}_{\mathrm{BC}}) = 2q f_\alpha(\mathbf{x}_{\mathrm{F1}}) + (1-2q) f_\alpha(\mathbf{x}_{\mathrm{F2}}), \qquad q < 0.5, \tag{22a}$$

$$f_{\mathrm{opp}(\alpha)}(\mathbf{x}_{\mathrm{BC}}) = \frac{1}{2q} f_\alpha(\mathbf{x}_{\mathrm{F1}}) + \frac{(2q-1)}{2q} f_{\mathrm{opp}(\alpha)}(\mathbf{x}_{\mathrm{F1}}), \qquad q \geq 0.5. \tag{22b}$$

In the above equations, $\mathrm{opp}(\alpha)$ is the lattice direction opposite to α, $f_\alpha(\mathbf{x}_{\mathrm{F1}})$ is the distribution function located in the first neighbour cell in the lattice direction α, while $f_\alpha(\mathbf{x}_{\mathrm{F2}})$ is located in the second neighbour fluid cell in the same direction.

The difference between our implementation and the original approach is that the estimated distribution function is originally imposed at $f_{\mathrm{opp}(\alpha)}(\mathbf{x}_{\mathrm{F1}}, t + \Delta t)$, in contrast to our case, which applies it to $f_{\mathrm{opp}(\alpha)}(\mathbf{x}_{\mathrm{BC}}, t)$. The subsequent streaming operation of the time step will transport it to the right fluid cell before the collision.

3.3 Imposing Macroscopic Variables in Ghost Cells

For the WALE model, central finite differences of the velocity field are needed to estimate the strain and rotation rates. In the case of the described inlet and outlet boundary conditions, applying Eq. (7) yields suitable macroscopic variables in the ghost cells. However, a bounce-back boundary condition, such as in Sect. 3.2, imposes only some of the distribution functions. Directly applying Eq. (7) would create questionable moment values that could dramatically affect the estimation of the eddy viscosity in the vicinity of the wall, resulting in inaccurate results.

To deal with this issue, we propose a new algorithm that is employed after the boundary condition and allows imposing the macroscopic variables without affecting the distribution functions that will be streamed to fluid cells and have been imposed by the boundary conditions in microscopic distribution functions. The idea is to alter the rest, i.e., the outward streaming distribution functions such that the evaluation of Eq. (7) will yield reasonable values. During this procedure, the algorithm checks the lattice directions in order to decide which of them point to interior fluid cells and are needed to impose the microscopic boundary conditions. These directions, denoted as i, are marked as *non-free*. Simultaneously, one can estimate *partial* density and velocity field as

$$\delta\rho = \sum_i f_i, \quad i \in \textit{non-free directions}, \tag{23a}$$

$$\delta\rho\mathbf{u} = \sum_i \mathbf{e}_i f_i, \quad i \in \textit{non-free directions}. \tag{23b}$$

We index the group of the *free* directions with j. Assuming that the groups of non-free and free directions have I and J elements, respectively, we have $n = I + J$, with $n = 19$ for instance for the D3Q19 model. The idea is to use the free directions to impose the four macroscopic variables, namely density ρ_0 and the three velocity components $\mathbf{u}_0$. In order for this algorithm to be functional, one needs to ensure that $J \geq 4$ for the ghost cell in question. Moreover, in order for the three components of the velocity to be specified, it must be ensured that for $\sum_j \mathbf{e}_j = (\alpha_1, \alpha_2, \alpha_3)$ the relations $\alpha_1, \alpha_2, \alpha_3 > 0$ hold true. In most scenarios, these two restrictions are satisfied and we also have $J > 4$, which results in the over-determined system

$$\rho_0 - \delta\rho = \sum_j f_j, \quad j \in \textit{free directions}, \tag{24a}$$

$$\rho_0\mathbf{u}_0 - \delta\rho\mathbf{u} = \sum_j \mathbf{e}_j f_j, \quad j \in \textit{free directions}. \tag{24b}$$

An efficient way to resolve this issue is the use of the equilibrium function Eq. (6) estimated by the imposed macroscopic quantities, ρ_0 and $\mathbf{u}_0$. In this way, we can reduce the number of unknowns to four. At this point, we ignore the distribution function of the zero lattice direction f_0, which will be used to satisfy the density ρ_0. The next step is to loop over the rest of the *free* directions, starting from the direction with the smaller α, and impose the equilibrium values until we have only three unknown distributions. We index the group of K equilibrium distributions by k, and obviously $n = I + K + 4$. In this way, we end up with a system of four equations with four unknowns.

Initially, we have to solve the linear system of the three equations, indexed m, originating from the first moment:

$$\rho_0\mathbf{u}_0 - \delta\rho\mathbf{u} - \sum_k \mathbf{e}_k f_k^{\text{eq}} = \sum_m \mathbf{e}_m f_m \implies \mathbf{b} = \mathbf{A}\mathbf{f}. \tag{25}$$

In the current implementation, an LU-decomposition is employed to solve the above linear system. The last step is the evaluation of f_0 as

$$\rho_0 - \delta\rho - \sum_k f_k^{\text{eq}} - \sum_m f_m = f_0. \tag{26}$$

It is important to mention that the ascending order during the step of the equilibrium functions is vital for the stability of the algorithm. In case that at least one of the members of the m group belongs to $\alpha \in [1, 6]$, the matrix $\mathbf{A}$ will be singular resulting in no available solution for the system of Eq. (25).

The proposed algorithm can also be applied straightforwardly to the D3Q27 stencil and can also be used in 2D with the D2Q9 stencil. As for the imposed velocity $\mathbf{u}_0$, as a first attempt and following a ghost-fluid approach, we use in this paper the interpolated velocity at the point $\mathbf{x}_{\text{BC}} + 2\varphi\mathbf{n}$. However, one could increase the accuracy, particularly in a turbulent boundary layer, by assuming the law of the wall in the normal direction and thus estimating the velocity components.

4 Results

To illustrate the capabilities of the new implementations in the AMROC-LBM solver, we present two benchmark cases, namely decaying homogeneous isotropic turbulence in a periodic box and the flow around a sphere at Reynolds number Re = 1000. The first test case serves the purpose of verifying the core LES models and investigating their interplay with the two available collision operators. The second test, on the other hand, verifies their integration with various boundary conditions, particularly embedded complex walls, and the AMR algorithm. Moreover, the proposed algorithm for imposing macroscopic variables in ghost cells will also be tested and evaluated.

4.1 Decaying Homogeneous Isotropic Turbulence

The numerical domain for the decaying homogeneous isotropic turbulence test case is a cube with a side length $L = 2\pi$. Periodic boundary conditions are applied at all sides. Assuming a uniform grid, this setup provides a unique and convenient way to test LES models without disturbances arising from physical boundary conditions or the resolution interfaces between levels of AMR.

Our initialisation of the flow field is based on the final saved iteration of a forced homogeneous isotropic turbulence case presented previously in [12, 13]. In this scenario, we restart the simulation in the AMROC-LBM solver but deactivate the force. Suitable local volume averaging is applied when creating the initial solutions for the lower resolutions. To ensure a fair comparison, all simulations in this chapter have been initialised based on a DNS with a resolution of 512^3 running with the REG SRT collision model. The reason for this choice is that the force has created slightly different Reynolds number flows for the case of STA and REG SRT, and thus a direct comparison of the curves would be difficult. The initial Reynolds number based on the integral length scale λ is 80. Moreover, a field arising from the regularised model can safely be assumed to be more accurate, and it can be expected that the effect of the initial solution will fade away over time. We have also restarted the STA SRT simulations from the non-regularised DNS of 512^3, and the results were found to be identical to the ones presented below.

In the plots in this chapter, we will compare two resolutions, namely 128^3 and 32^3 cells, for two turbulence models currently available in the AMROC-LBM solver. The first one is the WALE, which we want to verify and evaluate, and the second is the CSMA with $C_S = 0.1$. Simultaneously, we will compare the two aforementioned collision models, STA and REG SRT. Additionally, a DNS with the

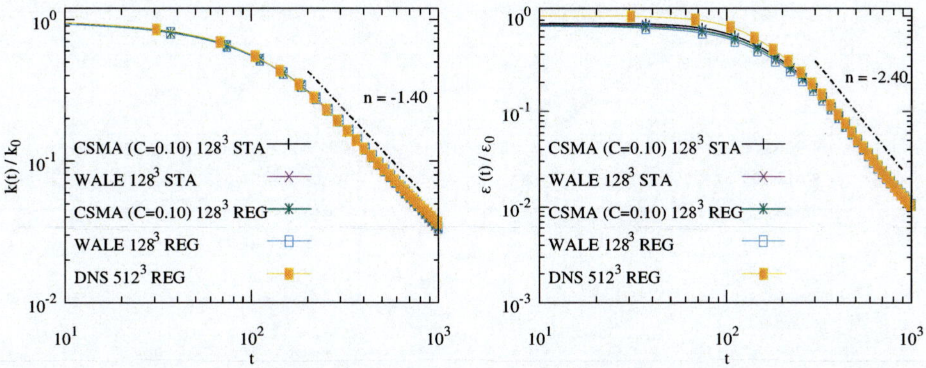

Fig. 2 Evolution of the turbulent kinetic energy k (left) and dissipation rate ε (right) for CSMA with $C_S = 0.1$ and WALE at a resolution of 128^3 cells for both STA and REG SRT. The DNS of 512^3 resolution with REG SRT has been added as a reference

REG SRT collision model with 512^3 cells will be shown as a reference solution. All simulations have run for a final time of 1000 time units.

Figure 2 presents the evolution of the turbulent kinetic energy k and dissipation rate ε for the resolution of 128^3 cells normalised by the initial data of the DNS of 512^3. The collapse of the LES curves with the reference DNS for the whole time in the case of the kinetic energy and for most of the time for the dissipation rate is imminent. The discrepancies appearing in the initial part for the dissipation rate are the effect of the local volume averaging resulting in fewer small eddies and thus a smaller initial value for ε. Therefore, examining these two plots, we cannot identify any differences between the two LES and the two collision models.

From the theory of decaying homogeneous isotropic turbulence, we expect that power laws of the type $k \sim (t + t_0)^{-n}$ and $\varepsilon \sim (t + t_0)^{-n-1}$ can describe the slopes in the current plots. In Fig. 2, we have also estimated the exponent $n = 1.4$, a value in the expected range in agreement with the literature [17].

Instantaneous 3D energy spectra and pressure fluctuation spectra at $t = 98.17$ time units are given in Fig. 3. Examining the energy spectra, their collapse in the energy-containing range is a strong proof that the LES models do not affect the large eddies, which is anticipated. Moreover, CSMA with $C_S = 0.1$ and WALE have produced identical results in the case of the same collision model, providing first evidence for the correctness of the WALE implementation. However, STA SRT seems to produce less dissipative results in the high wavenumber region, as it returns values closer to the DNS reference solution. This observation has also been recently reported by Nathen et al. [24].

Inspecting the pressure fluctuation spectra, one can notice that using the STA collision model the amount of small eddies has been considerably increased. It turns out that the departure of the energy spectra in the dissipation range for the two collision models coincides with this increase. Hence, we can speculate that the less dissipative behaviour of the STA SRT model is not because the small eddies carry

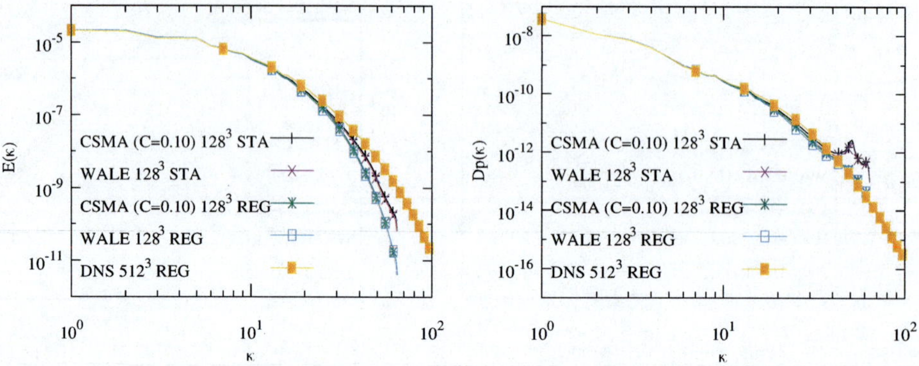

Fig. 3 Instantaneous energy spectra (left) and pressure fluctuation spectra (right) of the CSMA with $C_S = 0.1$ and the WALE for both STA and REG SRT for the resolution of 128^3 cells at $t = 98.17$ time units. The curves of the REG DNS on 512^3 cells are shown as a reference

more energy but because of an artificial rise in their numbers. Contrarily, the REG SRT model has estimated a solution much closer to the reference. We explain this observation with the fact that the regularisation procedure is constructed to impose the symmetries of the strain rate S_{ij} in the non-equilibrium part of the distribution function, Eq. (13), while reliably maintaining $\sum \mathbf{e} f^{\mathrm{neq}} = \sum f^{\mathrm{neq}} = 0$. The latter is not always guaranteed in the case of the STA SRT model [21], which can result in conservation errors in density and momentum, i.e., non-physical behaviour in the collision step. Such errors occur in particular for high Reynolds numbers or Mach numbers close to the LBM stability threshold.

To challenge the models more, Fig. 4 shows the evolution of k and ε for the case of the resolution of 32^3 cells. No combination of models is able to capture the reference curve of the kinetic energy in the initial phase exactly, although there are

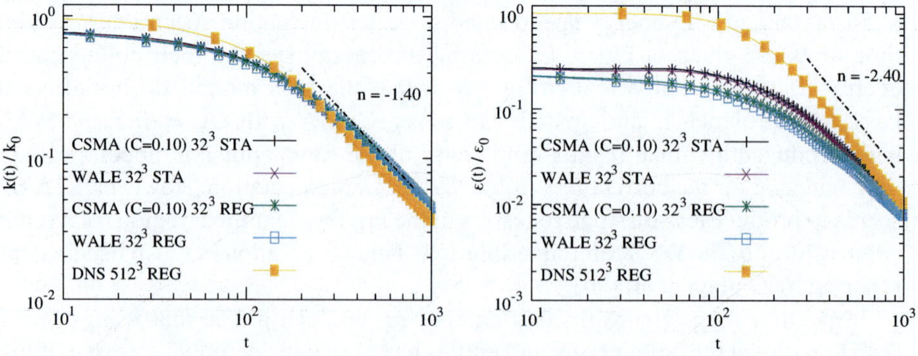

Fig. 4 Evolution of the turbulent kinetic energy k (left) and dissipation rate ε (right) for CSMA with $C_S = 0.1$ and WALE at a resolution of 32^3 cells for both STA and REG SRT. The DNS of 512^3 with REG SRT has been added as a reference

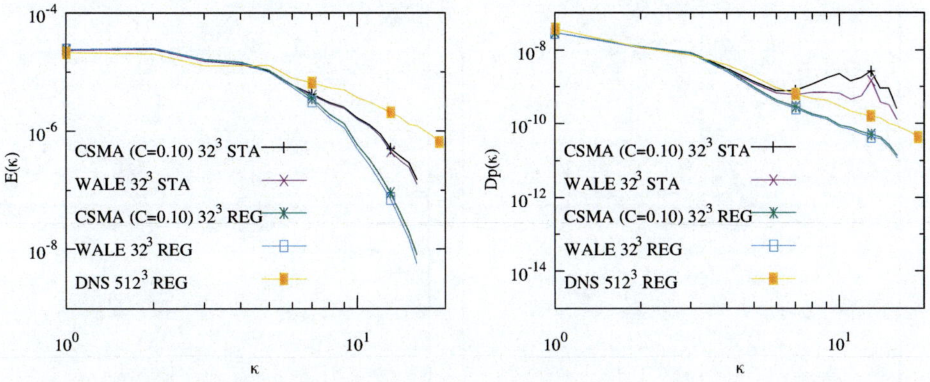

Fig. 5 Instantaneous energy spectra (left) and pressure fluctuation spectra (right) of the CSMA with $C_S = 0.1$ and the WALE for both STA and REG SRT for the resolution of 32^3 cells at $t = 98.17$. The curves of the REG DNS on 512^3 cells are shown as a reference

no evident discrepancies among them. The deviation from the DNS result is even larger for the dissipation rate, where only in the last time units there is a convergence of all the curves. The most interesting feature of this plot is the deviation of the curves of the STA and REG counterparts in the initial part of the simulation, with the former to show an observably less dissipative behaviour.

Following the same procedure, Fig. 5 presents the instantaneous 3D energy and pressure fluctuation spectra at $t = 98.17$. Due to the extreme coarsening of the grid, the energy of the big eddies has been slightly overestimated by both LES and collision models. Again, the STA SRT operator has returned less dissipative spectra in high wavenumbers. On the other hand, examining the pressure fluctuation spectra, the STA SRT model overestimates the small scales compared to the REG SRT model and the DNS. However, it is capable of following the DNS trend for more wavenumbers in contrast to the REG SRT that underpredicts the reference solution in the small scales.

Another important observation from the pressure fluctuation plot is the deviation of the WALE STA and CSMA STA models from the DNS in high wavenumbers. This deviation does not appear for the case of the REG collision model where the curves are identical. As mentioned previously, the CSMA estimates the strain rate locally based on the non-equilibrium part of the distribution function, while WALE applies finite differences. By reducing the resolution and thus increasing Δx and Δt, we have reduced the value of τ. This reduction leads to a higher value for the factor $\Delta t/\tau$ in the collision step, amplifying any inaccuracies arising from the imprecise evaluation of the non-equilibrium part in the case of the STA SRT operator. The estimation of the first-order moments, and thus the velocity components, is expected to be more accurate than the second-order moments, leading to an improved prediction of the strain rate based on a finite-difference stencil.

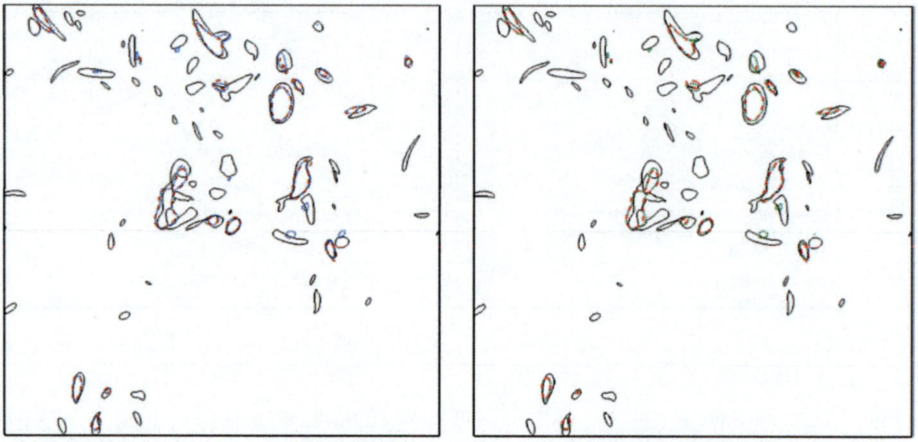

Fig. 6 Contours of vorticity magnitude ($|\omega| = 0.25$) at $t = 98.17$ time units, for the CSMA with $C_S = 0.1$ (left) STA (blue dotted) and REG (brown dashed) and for the WALE (right) STA (green dotted) and REG (red dashed) at a resolution of 128^3 cells. The black solid line is the DNS with a resolution of 512^3 given as a reference

Finally, Fig. 6 shows the instantaneous vorticity contours at $t = 98.17$ time units of the aforementioned models for the case of 128^3 cells. All combinations of models are able to capture the majority of the large eddies appearing in the reference solution of the DNS. Comparing the LES models, there are no apparent discrepancies, although the CSMA has produced slightly more small eddies. On the other hand, it is evident that the STA SRT model has predicted more small scales compared to the REG SRT, confirming our previous expectation from the pressure fluctuation plots.

4.2 *Sphere at Reynolds Number 1000*

To verify the coupling of the LES models with domain and embedded solid boundaries, and also with the AMR algorithm, the benchmark of flow around a sphere of diameter D at Reynolds number 1000 is selected. A computational domain of dimensions $[-2D, 6D]\times[-2D, 2D]\times[-2D, 2D]$ is used. The domain boundary conditions from Sect. 3.1 are applied, where an inlet boundary condition is imposed on the left side and outlet boundary conditions are applied at all other sides. The no-slip wall boundary condition on the body is modelled with the Bouzidi bounce-back condition, as sketched in Sect. 3.2. The mesh adaptation was set up to run with five levels in total with a refinement factor $r_l = 2$ for all levels. The mesh width of the coarsest grid is $\Delta x = D/20$. The scaled gradient [6] of the vorticity magnitude was chosen as a refinement indicator with a threshold value of 100. For the WALE model, we have also employed and tested the new algorithm for imposing

Fig. 7 Flow around a sphere at Re = 1000 simulated with CSMA with $C_S = 0.12$. Left: Isosurface of vorticity and planes coloured by velocity magnitude. Right: Computational mesh and distribution to processors, indicated by shading

macroscopic variables in ghost cells discussed in Sect. 3.3. The flow field has been initialised with the inlet velocity and an initial value for density.

Two simulations are discussed: one using the CSMA model with $C_S = 0.12$ and the other using the WALE model. Both computations use the REG SRT collision model. This decision was based on the superior behaviour of the REG SRT operator in the previous test case. The left plot of Fig. 7 visualises by colour the velocity magnitude for the case of CSMA in two planes and on an isosurface of vorticity magnitude for the value 100 at a time when the wake has been established. The shading in the right plot presents the distribution of the numerical domain to the employed processors and in addition the automatically refined mesh at the same time. It is evident that the refinement follows the isosurface closely. Figure 8 displays the two corresponding plots for the case of the WALE model. Comparing

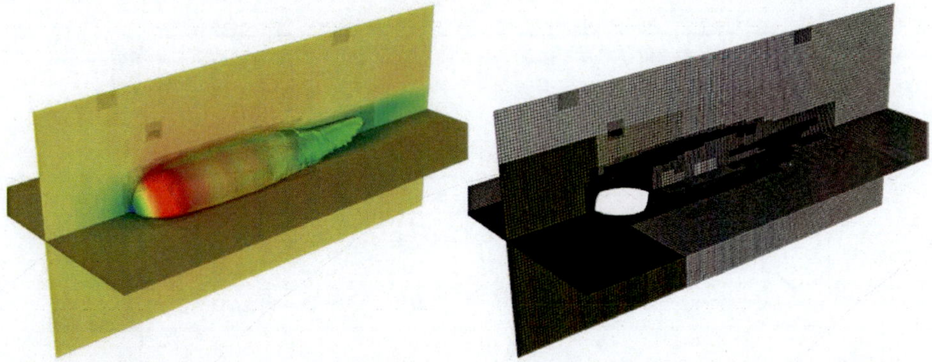

Fig. 8 Flow around a sphere at Re = 1000 simulated with WALE. Left: Isosurface of vorticity and planes coloured by velocity magnitude. Right: Computational mesh and distribution to processors, indicated by shading

the results of the two turbulence models, there are no significant discrepancies, and both computations exhibit a very similar 3D wake structure and according mesh refinement. Minor differences in both figures are due to the different nature of the two turbulence models. The CSMA tends to predict a more diffusive eddy viscosity field with higher values in the whole domain. This behaviour will invariably reduce the accuracy of the solution but simultaneously has the positive effect of stabilising numerical fluctuations, e.g., from boundary conditions. On the other hand, WALE tends to estimate lower eddy viscosity values in the majority of the domain but predicts larger values in and around fluid features. This distinction is the reason for any difference in the wakes and hence the dynamically adapted meshes.

To further enhance the comparison, Fig. 9 shows the vorticity magnitude for the two LES models in the xz-plane in logarithmic scale. Again, there are no significant differences between the shape and the shading for the two models. Examining the wake, particularly far away from the body, one can see some minor perturbations being emanated from the outlet boundary. The situation is slightly better for the CSMA model. Finally, Fig. 10 shows the evolution of the drag coefficients obtained during the last phase of the simulation for the two models. Both computations predict an average value of 0.461 that perfectly matches the theoretical value of 0.46 calculated from the standard drag curve for a sphere, and the value 0.464 estimated with a recently proposed formula [14]. The predicted value also matches the one reported in [27] and is very close to the value of 0.48 reported in [26].

At this point, we evaluate the algorithm for imposing macroscopic variables in ghost cells from Sect. 3.3. To do so, we have also run the simulation of WALE without invoking the new treatment. Figure 11 presents the enlarged normalised eddy viscosity fields, at the same time step, for the WALE simulations with and without the use of the new algorithm. The normalisation of the eddy viscosity is based on the value of the physical viscosity. Examining the magnitude, it is clear that the LES model was not triggered considerably in the vicinity of the wall. There are two reasons for this: First, the WALE model is constructed to reduce

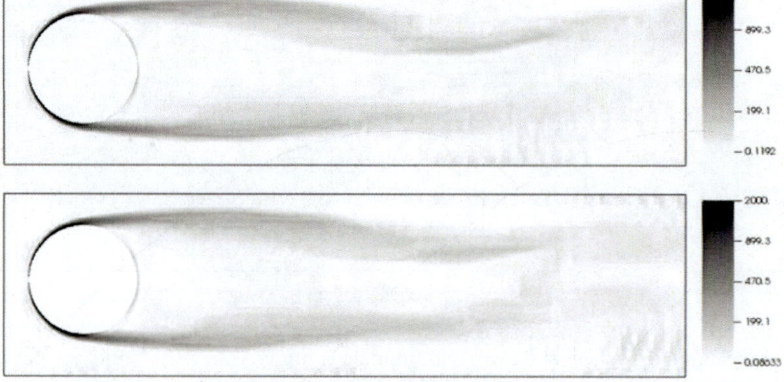

Fig. 9 Comparison of the vorticity magnitude between CSMA (top) and WALE models (bottom)

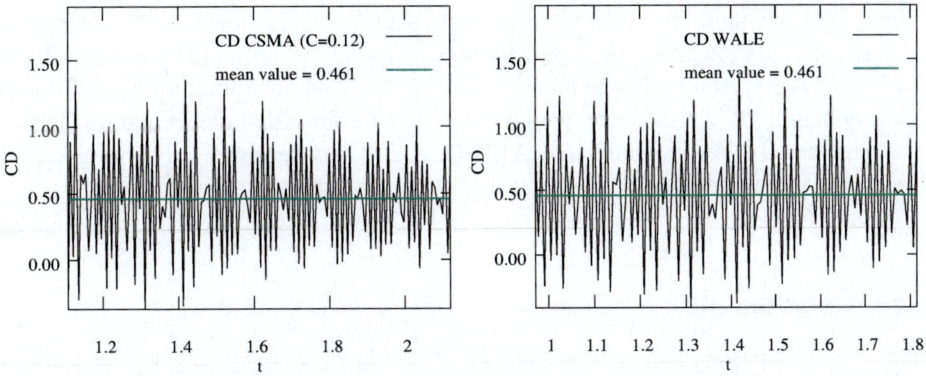

Fig. 10 Evolution of the drag coefficient CD for the CSMA model with $C_S = 0.12$ (left) and the WALE model (right). The dashed lines show the averaged values

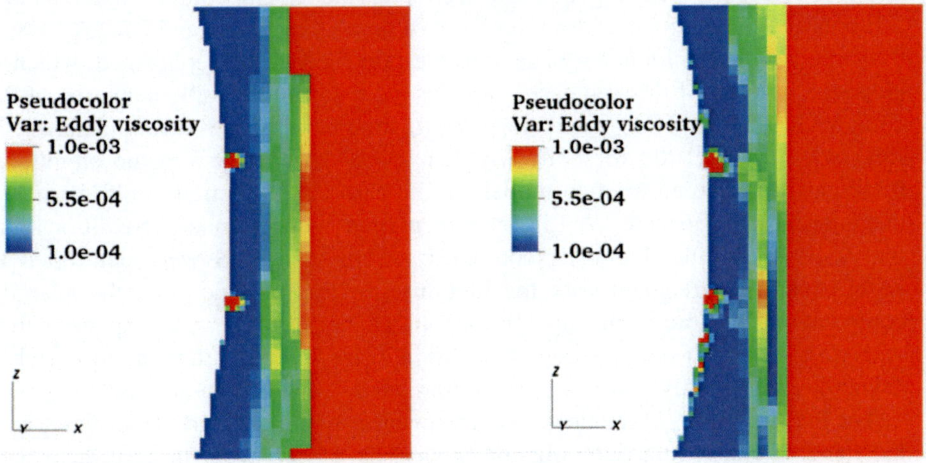

Fig. 11 Comparison of the enlarged normalised eddy viscosity with the application of the new algorithm for imposing macroscopic variables in ghost cells (left) and without (right)

the production of eddy viscosity close to the wall to represent the law of the wall more accurately. Secondly, in this specific case, the wall-near resolution is very high in order to approximate the curved geometry well and thus obtaining a very accurate estimation of the drag coefficient. The results confirm the consistency of the model in the case of a high-resolution mesh for a wall-resolved LES. Comparing the two graphics, one can conclude that the use of the algorithm from Sect. 3.3 has reduced the formation of a band of large eddy viscosity values close to the wall.

We close this section with a discussion of the performance of the AMROC-LBM solver for the sphere test case. Both simulations ran for 120 h wall time using 64 cores of 2.6-GHz Intel Sandybridge processors. The initial size of the adaptive mesh for both cases is ~8.7 M cells, while the final size is between 16 and 17 M cells. Application of the finest level with ~5.3 M cells is restricted to the vicinity of the body. On average, the CSMA simulation required 27.65 s per iteration, while the WALE simulation took 32.26 s, which corresponds to an added expense of only 17%. However, the final size of the mesh for the WALE simulation is also slightly larger than in the CSMA case, which altogether confirms that incorporating the more sophisticated WALE model into AMROC-LBM has resulted only in a moderate increase in computational time.

5 Conclusions

The main aim of this paper was to present the verification procedure of the WALE turbulence model recently implemented in our dynamically adaptive in-house lattice Boltzmann solver AMROC-LBM. The first step was simulating the test case of decaying homogeneous isotropic turbulence and comparing the energy and pressure fluctuation spectra with DNS of higher resolution and CSMA LES of the same resolution. Identical behaviour with the CSMA was confirmed, which—given the isotropy of this test case—verifies the algorithmic implementation of the core WALE model. The second step was to simulate the flow around a sphere at Reynolds number 1000 for both WALE and CSMA. There were no significant discrepancies between the two models in the case of the vorticity field, verifying the interplay of the new WALE implementation with boundary conditions and the AMR algorithm. The drag coefficients from both LES were confirmed to be in excellent agreement with the literature. In this specific case, the adaptive computation using the WALE model was found to be only 17% more expensive than with CSMA, which demonstrates that the increase in computational costs, when using the considerably more complex WALE model, can be kept modest.

For the simulation of the decaying homogeneous isotropic turbulence case, both the standard and regularised collision models have been used. In agreement with recent research studies, we have found that the energy spectra of the STA model are less dissipative. To further enhance the comparison, we have also presented pressure fluctuation spectra that highlight the fact that the STA collision model produces a large amount of small-scale perturbations, not present in the REG results. This behaviour of the STA collision model is unphysical and likely intrinsic to the model's handling of the non-equilibrium part of the distribution function. Moreover, in the case of the lowest resolution and STA collision model, we have shown that the CSMA model, which estimates the strain rate locally based on the non-equilibrium part, tends to enhance these instabilities compared to the WALE model, which uses finite-difference stencils of macroscopic variables.

Finally, a new LBM boundary condition construction algorithm for imposing macroscopic variables in addition to inward-directed microscopic distributions has been proposed. For instance, in the case of bounce-back wall boundary conditions, the resulting macroscopic moments are not well defined. By imposing suitable values, the finite-difference stencils can still be applied unaltered, hence yielding a plausible estimate for the eddy viscosity in the vicinity of the wall or other domain boundaries. A detailed analysis of the eddy viscosity from the WALE model close to the embedded wall, with and without the new algorithm, has confirmed the increased accuracy of our approach.

Acknowledgments This work was supported by UK Research and Innovation under grant EP/N509747/1 with project number 1831845. The authors also acknowledge the use of the IRIDIS High-Performance Computing Facility and associated support services at the University of Southampton.

References

1. Aidun, C.K., Clausen, J.R.: Lattice-Boltzmann method for complex flows. Annu. Rev. Fluid Mech. **42**(1), 439–472 (2010)
2. Berger, M., Colella, P.: Local adaptive mesh refinement for shock hydrodynamics. J. Comput. Phys. **82**, 64–84 (1988)
3. Bouzidi, M., Firdaouss, M., Lallemand, P.: Momentum transfer of a Boltzmann-lattice fluid with boundaries. Phys. Fluids **13**(11), 3452–3459 (2001)
4. Chen, H., Filippova, O., Hoch, J., Molvig, K., Shock, R., Teixeira, C., Zhang, R.: Grid refinement in lattice Boltzmann methods based on volumetric formulation. Phys. A **362**, 158–167 (2006)
5. Deiterding, R.: A parallel adaptive method for simulating shock-induced combustion with detailed chemical kinetics in complex domains. Comput. Struct. **87**, 769–783 (2009)
6. Deiterding, R.: Block-structured adaptive mesh refinement - theory, implementation and application. Eur. Ser. Appl. Ind. Math. Proc. **34**, 97–150 (2011)
7. Deiterding, R., Wood, S.L.: An adaptive lattice Boltzmann method for predicting wake fields behind wind turbines. In: Dillmann, A., Heller, G., Krämer, E., Wagner, C., Breitsamter, C. (eds.) New Results in Numerical and Experimental Fluid Mechanics X. Notes on Numerical Fluid Mechanics and Multidisciplinary Design, vol. 132, pp. 845–857. Springer, Berlin (2016)
8. Deiterding, R., Wood, S.L.: Predictive wind turbine simulation with an adaptive lattice Boltzmann method for moving boundaries. J. Phys. Conf. Ser. **753**, 082005 (2016)
9. Di Ilio, G., Chiappini, D., Ubertini, S., Bella, G., Succi, S.: Fluid flow around NACA 0012 airfoil at low-Reynolds numbers with hybrid lattice Boltzmann method. Comput. Fluids **166**, 200–208 (2018)
10. Feaster, J., Battaglia, F., Deiterding, R., Bayandor, J.: Validation of an adaptive meshing implementation of the lattice Boltzmann method for insect flight. In: Proceedings of the ASME 2016 Fluids Engineering Division Summer Meeting, pp. FEDSM2016–7782, V01AT12A007. ASME, New York (2016)
11. Feldhusen, K., Deiterding, R., Wagner, C.: A dynamically adaptive lattice Boltzmann method for thermal convection problems. J. Appl. Math. Comput. Sci. **26**, 735–747 (2016)
12. Gkoudesnes, C., Deiterding, R.: Evaluating the lattice Boltzmann method for large eddy simulation with dynamic sub-grid scale models. In: 11th International Symposium on Turbulence and Shear Flow Phenomena (2019)
13. Gkoudesnes, C., Deiterding, R.: Verification and validation of a lattice Boltzmann method coupled with complex sub-grid scale turbulence models. In: VI International Conference on Particle-based Methods - Fundamentals and Applications (2019)
14. Goossens, W.R.: Review of the empirical correlations for the drag coefficient of rigid spheres. Powder Technol. **352**, 350–359 (2019)
15. He, Y.L., Liu, Q., Li, Q., Tao, W.Q.: Lattice Boltzmann methods for single-phase and solid-liquid phase-change heat transfer in porous media: a review. Int. J. Heat Mass Transf. **129**, 160–197 (2019)

16. Hou, S., Sterling, J., Chen, S., Doolen, G.D.: A lattice Boltzmann subgrid model for high Reynolds number flows. In: Lawniczak, A.T., Kapral, R. (eds.) Pattern Formation and Lattice Gas Automata. Fields Institute Communications, vol. 6, pp. 151–166. American Mathematical Society, Providence (1996)
17. Huang, M., Leonard, A.: Power-law decay of homogeneous turbulence at low Reynolds numbers. Phys. Fluids **6**(11), 3765–3775 (1994)
18. Kin, N., Deiterding, R., Wagner, C.: High-resolution simulation of side flow past a generic model of a high-speed train. In: Dillmann, A., Heller, G., Krämer, E., Wagner, C., Breitsamter, C. (eds.) New Results in Numerical and Experimental Fluid Mechanics X. Notes on Numerical Fluid Mechanics and Multidisciplinary Design, vol. 132, pp. 421–431. Springer, Berlin (2016)
19. Krüger, T., Kusumaatmaja, H., Kuzmin, A., Shardt, O., Silva, G., Viggen, E.M.: The Lattice Boltzmann Method: Principles and Practice. Springer, Berlin (2016)
20. Laloglu, C., Deiterding, R.: Simulation of the flow around an oscillating cylinder with adaptive lattice Boltzmann methods. In: Ivanyi, B.H.V., Topping, P., Varady, G. (eds.) Proceedings of the 5th International Conference on Parallel, Distributed, Grid and Cloud Computing for Engineering. Civil-Comp Press (2017)
21. Latt, J., Chopard, B.: Lattice Boltzmann method with regularized pre-collision distribution functions. Math. Comput. Simul. **72**(2–6), 165–168 (2006)
22. Li, Q., Luo, K., Kang, Q., He, Y., Chen, Q., Liu, Q.: Lattice Boltzmann methods for multiphase flow and phase-change heat transfer. Prog. Energy Combust. Sci. **52**, 62–105 (2016)
23. Mauch, S.P.: Efficient algorithms for solving static Hamilton-Jacobi equations. Ph.D. Thesis, California Institute of Technology (2003)
24. Nathen, P., Gaudlitz, D., Krause, M.J., Adams, N.A.: On the stability and accuracy of the BGK, MRT and RLB Boltzmann schemes for the simulation of turbulent flows. Commun. Comput. Phys. **23**(3), 846–876 (2018)
25. Nicoud, F., Ducros, F.: Subgrid-scale stress modelling based on the square of the velocity gradient tensor. Flow Turbul. Combust. **62**, 183–200 (1999)
26. Ploumhans, P., Winckelmans, G., Salmon, J., Leonard, A., Warren, M.: Vortex methods for direct numerical simulation of three-dimensional bluff body flows: application to the sphere at Re=300, 500, and 1000. J. Comput. Phys. **178**(2), 427–463 (2002)
27. Poon, E.K.W., Iaccarino, G., Ooi, A.S.H., Giacobello, M.: Numerical studies of high Reynolds number flow past a stationary and rotating sphere. In: Seventh International Conference on CFD in the Minerals and Process Industries, p. 7 (2009)
28. Reyes Barraza, J.A., Deiterding, R.: A lattice Boltzmann method in generalized curvilinear coordinates. In: VI International Conference on Particle-based Methods - Fundamentals and Applications (2019)
29. Shao, W., Li, J.: Review of Lattice Boltzmann Method Applied to Computational Aeroacoustics. Archives Acoust. **44**(2), 24 (2019)
30. Smagorinsky, J.: General circulation experiments with the primitive equations, I: The basic experiment. Mon. Weather Rev. **91**(3), 99–164 (1963)
31. Succi, S.: The Lattice Boltzmann Equation for Fluid Dynamics and Beyond. In: Numerical Mathematics and Scientific Computation. OUP Oxford, Oxford (2001)
32. Tiwari, A., Vanka, S.P.: A ghost fluid lattice Boltzmann method for complex geometries. Int. J. Numer. Meth. Fluids **69**(2), 481–498 (2012)
33. Van den Akker, H.E.: Lattice Boltzmann simulations for multi-scale chemical engineering. Curr. Opin. Chem. Eng. **21**, 67–75 (2018)
34. Wang, J., Chen, L., Kang, Q., Rahman, S.S.: The lattice Boltzmann method for isothermal micro-gaseous flow and its application in shale gas flow: a review. Int. J. Heat Mass Transf. **95**, 94–108 (2016)
35. Xu, A., Shyy, W., Zhao, T.: Lattice Boltzmann modeling of transport phenomena in fuel cells and flow batteries. Acta Mech. Sin. **33**(3), 555–574 (2017)
36. Zhao-Li, G., Chu-Guang, Z., Bao-Chang, S.: Non-equilibrium extrapolation method for velocity and pressure boundary conditions in the lattice Boltzmann method. Chin. Phys. **11**(4), 366–374 (2002)